Praise for *Space 2069*

'It is rare to read something that so closely mixes science fiction with reality, but *Space 2069* does just that … [It] packs a sizeable punch … an intelligent portrait of where we may be in the next half-century.'

BBC Sky at Night

'Rich, topical and informative'

Physics World

'[A] skilful history of space exploration … A realist, Whitehouse emphasizes that, without a major breakthrough in rocket technology, travel to Mars will test the limits of human endurance and willingness to bear the expense. His forecast for 2069 is a struggling eighteen-man international base on Mars. China will have its own. A fine overview of the past and future of human space exploration.'

Kirkus Reviews

Praise for *Apollo 11*

'Terrific and enthralling'

New Scientist

'An authoritative account of Apollo 11 and the end of the space race, shedding light on the true drama behind the mission.'

The Observer

'Fascinating'

The Herald

'Fast-paced and tremendously readable ... What makes this book really stand out from other Apollo-based books is the inclusion of long quotes from interviews with astronauts such as John Glenn (the first American to orbit Earth), Eugene Cernan (the last man to walk on the Moon) and, of course, Neil Armstrong himself.'

BBC Sky at Night

'The book is at its most successful when Whitehouse gets out of the way of its protagonists, letting the astronauts and cosmonauts offer their own verbatim accounts of their often perilous – and occasionally fatal – missions. The real strength of this book is its tribute to the human qualities of these men – and they are all men, with the exception of the brief but gripping story of one female cosmonaut – who were willing to sacrifice so much.'

The Irish News

'In the most authoritative book ever written about Apollo, David Whitehouse reveals the true drama behind the mission, telling the story in the words of those who took part based around exclusive interviews with the key players ... [An] enthralling book.'

All About Space

'David Whitehouse's masterly narration of what he calls "the inside story" is profoundly gratifying.'

The Spectator

'Whitehouse has a reporter's gift for uncomplicated storytelling'

Financial Times

'One of the best books ever written about the lunar landing ... absolutely brilliant.'

Engineering and Technology

The Alien Perspective

BY THE SAME AUTHOR

Apollo 11: The Inside Story (Icon, 2019)

Space 2069: After Apollo: Back to the Moon, to Mars, and Beyond (Icon, 2021)

The Alien Perspective

A New View of Humanity and the Cosmos

DAVID WHITEHOUSE

This edition published in the UK and USA in
2022 by Icon Books Ltd, Omnibus Business
Centre, 39–41 North Road, London N7 9DP
email: info@iconbooks.com
www.iconbooks.com

ISBN: 978-183773-099-5

Typeset in Berkeley by Marie Doherty

Printed and bound by CPI Group (UK) Ltd,
Croydon, CR0 4YY

To Jill

'But the barriers of distance are crumbling;
one day we shall meet our equals,
or our masters, among the stars.'

ARTHUR C. CLARKE, *2001: A SPACE ODYSSEY*, 1968

'Far away, hidden from the eyes of daylight,
there are watchers in the skies.'

– EURIPIDES, *THE BACCHAE*, 406 BC

'Thoughts, silent thoughts,
of time and space and death.'

– WALT WHITMAN, 'PASSAGE TO INDIA', 1869

CONTENTS

ABOUT THE AUTHOR

Dr David Whitehouse is a former BBC science correspondent and editor. He studied astrophysics at the world-famous Jodrell Bank radio observatory under Sir Bernard Lovell. He is the author of several books, including most recently *Space 2069: After Apollo: Back to the Moon, to Mars, and Beyond* and *Apollo 11: The Inside Story*. He has written for many newspapers and magazines and regularly appears on TV and radio programmes. He has won many awards, including the very first Sir Arthur Clarke Award as well as the European Internet Journalist of the Year. Asteroid 4036 Whitehouse is named after him.

PREFACE

For me, the question of whether there is intelligent life in space is, alongside that about the existence of God, the most important question I know. After a lifetime of thinking about the subject, I was hesitant to write a book about it and initially prepared proposals for a book that skirted around the issue. It was my publishers, Icon Books, and my literary agent, Laura Susijn, who asked me questions and homed in on the essence of what I was stumbling towards. The question of whether there is intelligent life in space is a big one indeed.

During its writing, the project acquired several titles. One was *What If?*, which was very early on written on a folder containing my notes. There were so many questions, but I worried about a book that asked too many. I soon changed my mind because when it comes to looking for life, especially intelligent life in space, all we have are questions, and in asking them we illuminate the problems, our prejudices and our ignorance. At other times I joked that the book should be called *Everything We Know About Aliens* and should consist of 252 pages, all of them blank!

At one time I had in mind the forthcoming 50th anniversary of Carl Sagan's book, *The Cosmic Connection*, thinking that it was time to assess our cosmic perspective once more, as many others have successfully done, but the alien kept reappearing, and I thought more and more about the search for intelligent life in space and the gap left in the subject by Carl Sagan's absence.

I thought of demotions, of humanity not being the centre of creation, of the Earth not being the centre of the universe. I thought that finding intelligent aliens would be another in a long line of demotions of humankind, but as you will see I changed my mind, thinking that whatever or whoever is out there, if at all, we should not feel any the lesser for their existence.

And I worried about the contrast between the public interest in life in space and the relatively small number of people who have shaped the philosophy of looking for intelligent aliens, and I thought it was time for new voices, new ways of thinking and perhaps fewer bad jokes and brush-offs that cover up legitimate concerns.

Are we alone? There couldn't be a shorter yet more profound question, let alone one that strikes to the very core of what we are. What hubris, I thought, attempting to write a book about such questions.

Yet the question is there behind every corner on Earth, behind every ecosystem and living creature, behind every tenet of philosophy, every religious impulse, inside us all. Are there others? Is this wondering common in the cosmos? How much is parochial, how much universal?

Science is not driven by logic. It is driven by desire, a desire to find out, to experience something wonderful, bigger than ourselves and, in the case of aliens, beyond ourselves. It is strange that we should feel so passionate about beings we have never met. Or is it?

And so, a few thoughts and a few ideas about aliens and the place and prospects for life in the cosmos. If we ever encounter them, perhaps the thing we will have most in common are the questions.

I would like to thank Nick Booth for his help and insightful comments throughout this project, and Rebecca Charbonneau for showing me her doctoral thesis on the history of searching for intelligent life in space, which I thoroughly recommend. Thanks also to Carol Oliver of Macquarie University in Australia. I also thank Michael Rappenglück for discussions some time ago about caves and the cosmos.

I would like to thank Duncan Heath and James Lilford of Icon Books, my amazing literary agent Laura Susijn for all her support and of course my children Christopher, Lucy and Emily. My wife, Jill, has given me unwavering support and so many ideas. I cannot thank her enough.

David Whitehouse
Hampshire
2022

A QUESTION ASKED DIFFERENTLY

According to a recent survey, there are 540 objects within 33 light years of our Sun – next door in astronomical terms. This includes 373 stars, 88 brown dwarfs, which are failed stars, 21 white dwarfs, which are dense remnants of dead stars, and 77 planets that circle some of the nearby stars. Most of the stars are red dwarfs that are smaller and dimmer than our Sun but there is a scattering of stars like our Sun. A radio signal sent from Earth in the late 1980s would by now have reached all of them. Because most of the stars that comprise our constellations are much further away, many of our constellations would be recognisable in the sky as seen from these nearby stars and worlds. The view from the nearest one, a red dwarf called Proxima Centauri, which has at least one planet orbiting it, would have an additional star in the constellation we call Cassiopeia. This is our Sun. In a way, to any inhabitants of that system we are Cassiopeians. They might look at our Sun from time to time, but if they judged it by its rather average characteristics they might not pay it much attention. But if they had the ability to detect radio emissions then perhaps they might take a second look.

There is a greater number of stars, 2,034 to be precise, that have a unique view of our planet, even from the greater depths of space beyond Proxima Centauri. These are stars which currently or in the not too-distant past or future could detect our

Earth passing across the face of our Sun. If there are aliens on any planets orbiting those stars with at least our level of technology, they would be able to spot our world. Should they have telescopes only slightly more advanced than ours, they would be able scrutinise the atmosphere of our tiny rocky world as it transits our home star. They would find the atmosphere unusual: a lot of oxygen, which in the absence of life wouldn't last long, and perhaps looking along the spectrum of light from this planet a hint that something on its surface might be absorbing light from its star, perhaps powering a metabolism and by implication an ecology.

Perhaps it, they or them might send us a radio signal or a laser flash. Perhaps they might dispatch a probe knowing it will take hundreds, possibly thousands, of years to get here, and perhaps that is trivial to them. Or perhaps there is nothing, and all those worlds and voids are lifeless.

How do we prepare ourselves for contact with aliens? We have only our logic, our assumptions, our emotions and especially our fears to prepare us. It has been said that until we have alien contact, humanity will be blind and afterwards it will be too late.

Our own planet is the only place we know of where there is life. Planets are not the only places where life could exist and possibly not the main place. Where else life could arise may surprise you, and I do not mean the much-talked-about prospects for life in our own planetary system.

Of these there is of course Mars, for some the best hope for nearby life. Perhaps life arose independently there, or perhaps life on Earth and possibly Mars are related in some way, as we shall see. We have sent more than 30 missions to this world, many carrying sophisticated instruments on rovers to explore its surface. We see a world we recognise: deserts of sand, boulders, dried-up river beds, sand dunes blown by a thin wind and a peach-coloured sky. But we haven't found life.

Beneath the icy surface of Jupiter's moons Ganymede, Europa and Callisto there is water. Protected for perhaps hundreds of millions of years, these oceans would have access to nutrients and possibly have hydrothermal vents, like the ones on the sea floors of Earth around which so much strange life congregates. Saturn's moons Rhea and Enceladus are the same, and like Europa, they have recently been found to have explosive jets of water vapour that erupt into space through fractures in their crusts known as 'tiger stripes'. Saturn's major moon is Titan, which has a chemically rich, thick atmosphere and oceans of liquid hydrocarbons. Even further out in our solar system, Uranus' moons Titania and Oberon, and Neptune's large moon Triton, also possibly have sub-surface oceans. The dwarf planets Pluto and Eris may also possess similar oceans, as may the hundreds or thousands of the distant members of the Kuiper Belt. Each may harbour life.

We have our imagination. The *Dune* novels have their stars and planets. The fictional Arrakis is described as the third planet around the bright star Canopus in the southern hemisphere constellation of Carina the Keel, itself part of the great ship Argo Navis that took the Argonauts on their mythical quest for the Golden Fleece. Buzzell is a water world populated only by tiny islands. There is the forest world of Endor in *Return of the Jedi*, the overpopulated Coruscant seen in several *Star Wars* films, Bespin with no solid surface in *The Empire Strikes*

Back and the iconic and mysterious desert world of Tatooine, Luke Skywalker's childhood home. *Star Trek* has the United Federation of Planets, and in the year 2267 Captain James Tiberius Kirk said that humanity was on a 'thousand planets and spreading out'. The Federation promises a hopeful future, but we also imagine the ruthless Xenomorph of *Alien* and the marauders of *Independence Day*.

What does an alien look like? How would it or they think? What is their perspective on, well, life, the universe and everything? How would we make contact with them? And what exactly is life? Would we recognise it if we saw it?

We have only our imagination, logic and emotions to guide us when contemplating these questions and all of those are flawed if not misleading. But we have no choice, they are all we have. We need scientists and artists, philosophers and poets and perhaps clichés like those of *Star Trek*, *Star Wars*, *Alien*, *Independence Day* and *Close Encounters of the Third Kind*, however enjoyably unbelievable they might be. By the way, *The War of the Worlds* won't work, the biology in *Alien* is impossible and the alien at the end of *Close Encounters* is far too human-looking.

Would aliens arrive and say to us, as Klaatu the humanoid did in the 1951 film *The Day the Earth Stood Still*, 'If you threaten to extend your violence, this Earth of yours will be reduced to a burned-out cinder. Your choice is simple: join us and live in peace or pursue your present course and face obliteration.' No they wouldn't. That is science fiction, naïve and psychological. Our speculations about aliens illuminate much about the cosmos and humans, but aliens are aliens. How can we ever know them?

We once thought we were the centre of all things and that the universe was ours, in fact made for us. The Sun circled the Earth as did everything else and God was intimately involved in our affairs. There was a great chain of being that linked us and our world to heaven and eternity. But now we know differently. The Earth circles the Sun, which is an average star, one of billions in our galaxy, which itself is one of billions in the universe. Although we have yet to find an analogue of Earth, we know that planets are plentiful. Our origin and environment seem commonplace. Everything we've ever learned about the universe tells us that we are nothing special, except for one thing.

That one thing is you. Intelligent life, indeed, life of any sort. As far as we know, it only exists here on our planet circling our ordinary star. Humanity has been demoted many times – from the centre of the universe, from a supernatural creation, from living in an unusual star system. But our status as the only life we know of in the cosmos remains. We are special, unique, nobody can yet show otherwise.

But for how long? How would we react if we found out we were not alone? It could happen tomorrow or never. Some believe it already has. Suddenly humanity would be incomplete, not the pinnacle any more but a subset of what is possible. Finding out we are not alone, either in terms of a signal, an artefact or even a visitation, would not just tell us there is someone or something else, it would tell us that the universe is teeming with life, for as we have acknowledged there is nothing special about our insignificant corner of it.

We are passing through a crucial historical threshold. Within the past few decades we have arrived at a point where we could contact aliens. Many expect to do so, some are surprised we haven't already done so. Where are they? What are they like? What is their history? What is their outlook? Do they know about us? The first stars appeared over 10 billion years ago and the Earth is 4.6 billion years old, so perhaps the first intelligent species appeared billions of years ago. Imagine aliens older than our planet.

Think of that, a universe full of life, many different ways of being. We would ask ourselves, what is our place among them? Does it change our future? Does it make it better, hopeful or perhaps introduce a new fearfulness as we cower in the face of what could be an incomprehensible cosmic power? We should think about these things while we can, as the story of humanity might take a turn into the unknown when even our dreams are changed, or one might say, invaded.

Einstein said that the experience of the mysterious was the most beautiful thing and the source of all truth in art and science. He went on to say that he to whom the emotion of the mysterious is a stranger, who can no longer pause to wonder and stand wrapped in awe, is as good as dead – his eyes are closed. I will add to that. We must face the experience of the alien, a different creature's views and philosophy, intelligent but without our human mentality, even if this is perhaps beyond our imagination. Immortal possibly and, as far as we could appreciate, maybe almost omnipotent, transcending our history, our morals and our religions. They would not see the universe the way we do. As the American physicist Robert Oppenheimer said: 'In another world the basic questions may have been asked differently.'

Everything could be different from the alien perspective.

THE BIG SKY

On the beach at night,
Stands a child with her father.
Watching the east, the autumn sky.

– WALT WHITMAN, 'ON THE BEACH AT NIGHT', 1871

Walt Whitman's poem 'On the Beach at Night' is about a child holding her father's hand. They are looking at the night sky, at the ascent of the lordstar Jupiter and higher to 'the delicate sisters the Pleiades', as he puts it. But the clouds are coming, and the little girl silently weeps as one by one they obscure the stars. Weep not child, says her father, for they shall not long possess the sky. Watch again another night and the Pleiades will return. All those stars both silvery and golden, he adds, shall shine again. But then he does something unexpected. I imagine he crouches down to put his head next to hers and says softly, 'Mournest thou only for Jupiter? Considerest thou alone the burial of the stars.'

The stars are magnificent but temporary and I wonder if it is the same for life. There will come a time when all the stars are gone, every last one, when there will be no more hopeful sunrises, no more beautiful sunsets, never again a perfect day. Just as there must have been a first star to end the early cosmic dark ages there will be a last, and when it goes out then the cosmic age of light will be over, occupying an ever-smaller sliver of the receding past. As we contemplate our universe, we are beginning to see endings and how things will fall apart. In our story, we will meet those twin undertakers infinity and entropy who do not wait for us at the end of time. They are here, they always have been. There is an inscription on the tomb of Tutankhamun that could have been written by them: 'I have seen yesterday, I know tomorrow.'

The comfort that the father gives his daughter – that the clouds will not devour the stars – is true in her lifetime and for far beyond the lifetime of the Earth. But he then takes away that temporary respite when he talks of the very end of those stars. Perhaps he then feels he has gone too far, so he whispers, 'Something there is more immortal even than the stars.' As we shall see, there are many things more immortal than the stars, but will life be one of them? Especially conscious life that can foresee how the universe will change as it ages and how it must change to survive? Can life ever secure a beachhead on eternity's shoreline, or is its fate in the universe one of desolation, a brief tale wrecked on the rocks of ruined power and splintered space time?

I wonder if some intelligence will occupy a future shore 10 trillion years hence – a frozen beach that aeons ago lost the warmth needed to save it. It may look at a sky of eternal dark with nothing visible beyond its own desolated galaxy. Everything else will have been pulled beyond the observable horizon by the expansion of the universe. Will it, wretched and unhappy perhaps, wonder what it used to be like when life was easier? Will it look to the future with optimism and joy? Will the unanswered questions loom larger then?

How do we know this? How can we imagine trillions of years into the future? The information we have obtained about the universe in just over 400 years since the invention of the telescope has been remarkable even if it has torn humankind away from a place in paradise.

By a long way, most of the energy that reaches the Earth from the cosmos comes from the Sun, which gives the Earth 170 petawatts (a petawatt is 1,000,000,000,000,000 watts) of radiant energy – the power for our planet. The energy that arrives from the rest of the universe is miniscule in comparison, but we have used it well, and more is gathered every day.

At any dusk in northern Chile, the Paranal Observatory, some 2,635 metres above sea level, swings into action. Its four large telescopes are flanked by a coordinated array of smaller ones. They follow a laser beam projected into the sky that allows their adaptive optics to partly compensate for the distortion introduced by the turbulence of the atmosphere. The telescopes are so powerful that they could see the headlights of a car on the Moon and objects 4,000 million times fainter than those that can be seen with the unaided eye. Tonight, it is looking at star formation regions in the Tarantula Nebula – one of the most spectacular star formation regions in the solar neighbourhood. Not far away, the ALMA (Atacama Large Millimeter Array) of 66 interlinked radio telescopes is looking for radiation from a trio of planets orbiting an infant star.

On mainland USA, the main radio telescope at Green Bank Observatory in West Virginia – the world's largest fully steerable radio telescope at 100 metres across – is looking for molecules in the Orion Nebula, the nearest stellar nursery to Earth. We shall come to know Green Bank well. Further west at Kitt Peak, Tucson, the Mayall Telescope is gathering data to assemble the largest 3D map of the universe ever made, and in California the historic giant Mount Palomar 200-inch Hale Telescope is looking for stars that vary in brightness among the dust and gas lanes in the outer regions of a distant galaxy.

The Keck Observatory in Hawaii is looking at several bizarre objects near the galactic centre. Their true nature is difficult to discern as they are mostly hidden behind tracts of interstellar dust, but analysis of their light suggests they are moving extremely fast around a supermassive black hole that will one day devour them. Also on Hawaii, the Gemini North Telescope is looking for the light of planets orbiting nearby stars, and the nearby Subaru Telescope peers at the environment of quasars – supermassive black holes devouring stars, gas and dust partially obscured in a newborn galaxy visible from near the edge of the observable universe. The James Clerk Maxwell Telescope, another instrument placed in this atmospherically favourable area, is looking at how magnetic fields influence the birth of stars.

China, with its newly built record-breaking radio telescope, is sifting the cosmic static for signs of a signal from intelligent life. At the same time, as radio telescopes are not limited to night-time observing, the Effelsberg Radio Telescope in Germany is scrutinising a jet of superheated ionised gas propelled out into space by a massive black hole. The e-MERLIN array of telescopes centred on Jodrell Bank in the UK is joining in.

In underground chambers there are sub-atomic particle detectors that are so sensitive they need shielding in, for instance, a cavern built beneath the Alps. Scientists hope they might detect the passage of such a sub-atomic particle that actually comprises most of the universe. At the South Pole, detectors frozen into a square kilometre of ice look for ghostly particles from exploding stars or from the dawn of time.

In space, a flotilla of spacecraft continues the watch. The Hubble Space Telescope is undertaking an ultraviolet light survey of nearby galaxies and looking for the furthest star ever seen. The Transiting Exoplanet Survey Satellite looks for dips in the brightness of stars due to planets moving in front of them. Another space observatory looks out for flashes of gamma rays from the hot and violent parts of the cosmos. The James Webb Space Telescope peers back to the first stars ever formed.

Sentinel satellites keep a watch on the Sun looking for the first signs of solar flare that might in a few days' time cause disruption on Earth. Passing over the Moon's south pole, a spacecraft is making observations of ice scattered in the lunar dust. Data is also sent back from spacecraft on the surface of Mars and orbiting it. Information from Jupiter comes in, as it does from mankind's most distant emissaries – Voyager 1 and 2. Voyager 1 is the most distant object we have ever sent into space. It's over 14 billion miles away. Light, which can reach the Moon in just over a second, takes 21 hours and 25 minutes to get back from it.

We have learned so much from so little. Some years ago, there was a dinner held for astronomers and at each place setting was a card that read: 'Not to be turned over until the meal has commenced.' On the other side of the card was the message: 'In turning over this card you have expended more energy than that received by all the telescopes ever turned towards the sky.' That tiny amount of energy has taught us so much.

Humans have emerged silhouetted against the cosmos as much a part of it as the stars and the black holes we observe. Everything about us reflects the history of our universe. Subtle, seemingly random events 13.7 billion years ago during the first billionth of a second after the Big Bang have shaped us as they have the structure of atoms and logic, the nature of numbers and chaos, the flow of time. All have found their way inside us in a great cosmic connection that few are aware of. In our daily lives and travails, we do not appreciate that each of us is a masterwork – an expression of a fundamental and perhaps purposeless logic. Our atoms come from the Big Bang and the stars. We would not be here but for a puzzling alignment of electrons in two elements that allow stars to exist. Our DNA and form are intertwined with our planet and its ecology. In the future it will weave itself even tighter into the cosmos. But will something of the essence of us survive into the countless aeons to come? A faint distant cry of what we used to be or wanted to become. And will there be cries from others born under the light of alien suns or escapees from altered flows of time that will join ours in a desperate chorus for long-term survival in a changing universe?

I often stare at the Hubble Space Telescope's Ultra Deep Field images. One is a small region of space in the constellation of Fornax – the Furnace. Hubble looked intently at a patch a tenth of the apparent diameter of the full Moon as seen from Earth and produced the deepest image of the universe ever taken. There are over 10,000 objects, most of which are galaxies. Scientifically, it shows high rates of star formation during the very early stages of galaxy development less than a billion years after the Big Bang. These young galaxies are smaller than ones we see today. Ten years after it was first obtained, Hubble scientists homed in on a section of the Hubble Deep Field to take an even closer look. This was named the Hubble eXtreme Deep Field. Here there are about 5,500 galaxies, the oldest of which are seen as they were 13.2 billion years ago.

I get lost in this image, and in the subsequent James Webb Space Telescope's Deep Field, in their galaxies of many shapes and colours: bright ones, so-called spirals and ellipticals, some distorted by gravitational interactions with each other, others faint and mysterious at the limits of detection. Each one has its own stars, lanes of gas and dust, stellar nurseries, the wreckage of dead stars – hundreds of thousands of millions of them, all there when the universe was a fraction of its current age. There are streams of galaxies outlining giant space voids like the surface of a bubble. Some of them are crackling with newborn stars full of potential. Others have the dull red leer of ageing stars, ancient tales and tragedies. All this in just a handful of the 2 trillion galaxies in the known universe. And planets? From what we now know about the frequency of planets there are billions of planets in this image, every second there are countless sunrises and sunsets, countless opportunities and stories. The poet Muriel Rukeyser wrote that the universe is made of stories, not of atoms.

What of the stories of life, of the many ways of living? If life began there, was it stillborn on millions of worlds? Did it flounder on many more, flourish on some? The stories that make up the cosmos have been told countless times in each of those immense blurs of light, civilisations like ours in the outer rim of galaxies, federations in their crowded central regions, biological beings, mechanical ones and who knows what else.

The galaxies of the Deep Field images are not like that now. They will have evolved, grown larger, while many of their stars will have died or changed. To see what they would have become we just have to look around us. In those long-gone galaxies there were billions of stellar nurseries bequeathing a dusting of bright young stars. Not long, cosmically speaking, after these galaxies, there was one unremarkable and indistinguishable among all the rest, which in one of its spiral arms birthed a cluster of stars, one of which was our own Sun.

We stand at the threshold of a transformative age in astronomy and space science – a time when we plan telescopes beyond the imaginations of previous generations; the incredible amounts of data collected by these instruments will be crunched not by ourselves but by artificial intelligences. As we peer deeper into the cosmos and contemplate our first voyages to the stars, we also contemplate others and our ties to them that may increase our connection to the cosmos and enlarge our conception of experience.

This book is about us and the others we might share the cosmos with; why and how we must look for them, and what it would mean if we found them or if they found us. But it is about more than that. It is about our common future and the most difficult thing one can contemplate in the universe: long-term survival.

The girl in Walt Whitman's poem was introduced to the fact that the universe changes. It was so very different at the beginning to as it was just before the first stars were formed, and it will be again very different when they are all gone. It is said, mistakenly in my view, that the beauty of all things is that they must end. Jean-Paul Sartre said that life is drained of meaning when the illusion of life being eternal is lost. But is there any beauty to be found when the time comes when everything has nothing to do and nowhere to go?

Eventually I will take you to a place to show you what I mean. It's a place we will come to know well, though it is by no means the end of the journey we shall undertake. It is a place where time is older than you can imagine and where everything you ever knew is gone, left far behind as an infinitesimal flash growing ever more distant before the almost countless subsequent aeons of darkness. If you are drawn here by the eternal questions, you will be disappointed for there is nothing here for you, or anyone, or anything. Time and space are almost irrelevant.

But first we have to stay in the here and now and look around us.

JIG OF LIFE

'"So deep is the conviction that there must be life out there beyond the dark," he says. "One thinks that if they are more advanced than ourselves they may come across space at any moment, perhaps in our generation. Later, contemplating the infinity of time, one wonders if perchance their messages came long ago, hurtling into the swamp muck of the steaming coal forests, the bright projectile clambered over by hissing reptiles, and the delicate instruments running mindlessly down with no report."'

– LOREN EISELEY, *THE IMMENSE JOURNEY*, 1957

'In the case of every other predominant animal the world has ever seen … the hour of its ascendancy has been the eve of its entire overthrow.'

– H.G. WELLS, 'THE EXTINCTION OF MAN', 1897

The 19th century Scottish philosopher Thomas Carlyle said that the universe was a sad spectacle: 'If they [the stars] be inhabited, what a scope for misery and folly. If they be not inhabited, what a waste of space.'

If you look at images of distant clusters of galaxies knowing that each galaxy is 100,000 million or so stars, or at images of much closer starfields where there seem to be so many stars like grains of sand on a beach, Carlyle's sentiment is understandable. There is such a lot of space out there, more than is imaginable to us. But then I look at life on Earth, and I disagree with Carlyle. If there were no life found anywhere in the cosmos, not around any of those stars or in any of those galaxies, other than that on Earth, then it would not be a waste of space, not a tragedy. Life on our own planet is glorious in its designs, diversity and abundance. On our world, everywhere that it is possible for life to be we find it, and our ideas about what is possible for life are changing all the time. If the universe offered up only one place for life, then the Earth has done it justice. If you look to the skies, to those stars and galaxies, you may find life and intelligence, but I contend that nothing out there will detract from the magnificence of what you see here, of what you and I are a part of. Made from the same atoms as stars, on this world nature has woven a magnificent tapestry, and its threads are time and death.

Any thought of there being life elsewhere in space, on the various planets and moons of our own planetary system, on worlds circling other stars and beyond, must consider what is here as a reference. Life as we know it is our starting point.

Life emerged from a dead cosmos initially composed of 99.9 per cent hydrogen and helium – the two simplest atoms – out of which you can make, as far as we know, very little except for simple stars which formed in what has been called the 'cosmic dawn' 250 to 350 million years after the Big Bang. All atoms except hydrogen, helium and a little lithium have been made inside subsequent generations of stars, built from the wreckage left over when some of these first stars exploded. These stars had internal temperatures and pressures great enough to push lighter atoms together to form heavier ones with a little release of energy that emerges as starlight. Life as we know it requires more complex interactions between atoms than hydrogen and helium can achieve. We are made of carbon, nitrogen and oxygen and so much more. For Earth-type life, carbon is the key as it has the ability to form chains, making the range of molecules that can be made using it very large. On our planet, relatively few atoms, seventeen, are essential to life and they are those elements that are most commonly found in the Earth's crust. Aluminium and silicon, although very common, are conspicuously rare in the human body. Silicon, with its ability second only to carbon to form chains, seems necessary for plants but not mammals.

Earth life has exploited all sources of available energy, including other forms of Earth life. It has a range of senses – eyes, sonar, electric and magnetic senses – and the intelligence to use them. Flight has been developed many times independently by birds, bats, fish, insects and the extinct pterosaurs. Eyes of varying designs have also been invented separately by scallops, snails, octopuses, vertebrates, insects and other arthropods. The process of photosynthesis – extracting energy from sunlight – has also been created independently numerous times, for example, purple bacteria use a very different molecular membrane system from the chlorophyll method used by plants, and other organisms use sulphur reactions to get energy from light.

Many creatures have learned how to use tools, not just ourselves but apes, birds, otters, elephants and dolphins. Some believe that octopuses, dolphins and perhaps even some birds may have cognitive abilities approaching our own, but we are the only organisms who have learned how to build wheels, telescopes, computers and molecular biological engineering.

Only one form of life on Earth has learned how to contemplate its place in the universe.

All around us is an incomplete picture of the life that came before us. Fossils are a selective window on past life. Organic material replaced by rock imprints show the shapes of creatures with all other information about them being lost. We have these fossils, footprints and even fossilised dung – coprolite – but there are vast tracts of the history of our planet about which we know very little. Prior to the Cambrian era – 543 to 490 million years ago – when life was predominantly oceanic, creatures had no hard bodies or skeletons to be fossilised. There are some fossilised bacterial remains from the earlier Proterozoic era – 2.5 billion to 543 million years ago – and towards its latter end when the Earth was frozen there are indications of the existence of worms and feathery-looking creatures. But there are only microfossils from earlier epochs.

We have catalogued what we have discovered, dividing it into various species, giving each a genus, family, order, class, phylum, kingdom and domain in a hierarchy that illuminates the way life works and perhaps gives an indication of what could happen elsewhere. As part of our curation of life on our planet there are scattered across it many museums of natural history that hold a portion of Earth's wonders. Let us take them a step further.

Walk with me through some imaginary near-future virtual Museum of Life on Earth where all the life forms of our planet, as far as they are known, both present and extinct, are gathered. I wonder, if we were able to show such a museum to a creature born under the light of a different star, how much would we have in common and how much would be parochial to Earth? I'm reminded of the first European biologists who explored South America and found animals in familiar ecological niches performing similar functions to those seen in their homeland, but the creatures were strange and a little alien to them. The science fiction writer and author of *2001: A Space Odyssey*, Arthur C. Clarke, once wrote that 'nowhere in space will we rest our eyes upon the familiar shapes of trees and plants, or any of the animals that share our world'. That may be so, but I suspect it may not.

The entrance is an archway brightly coloured by gigantic fish fins. Mantis shrimps build elaborate tunnels and decorate the entrances with such fins. Here they beckon you to explore the wonders of Earth life.

My museum has a central spectacula with various wings, galleries and precincts devoted to different domains and kingdoms of life. As you enter, high above you are changing displays highlighting this diversity. As you walk in on this occasion, you are bathed in the dim light of the ocean's twilight zone a few metres beneath the surface as a sea angel flaps its tiny 'wings', accompanied by smaller so-called sea butterflies. They seek out food, and as your eyes become accustomed to the gloom you begin to make out rich red and peach colours that are often displayed by such pteropods. They are interrupted by something of a vastly different scale – the largest animal who has ever lived on planet Earth – a 190-tonne blue whale swimming past in search of shoals of krill. Life on Earth occupies many scales, from the blue whale to the smallest life forms just 8.5 thousandths of a millimetre in size when fully grown. Before you depart the dim reaches of the ocean, other creatures become apparent. Here come the parasites found in all ecosystems. Marine bacteriophages (viruses that attack bacteria) kill about 5 per cent of the tiny photosynthetic plankton in the ocean every single day that provide about 20 per cent of Earth's photosynthesis harvest.

Some animals have what we call culture. Sperm whales – the creatures with the largest brains on Earth – live in groups. There are regions of the world, such as off the coast of British Columbia, where adjacent groups of whales that are genetically identical behave differently, such as employing differing hunting and feeding strategies with different diets that they learned from their parents. So many animals behave in ways we are only just beginning to appreciate.

Consciousness, that undefinable ability that Virginia Woolf called 'a wave in the mind', appears to have evolved several times, which suggests it must provide important benefits when it comes to survival. One of them might be behavioural flexibility, allowing an organism not to react in an automatic reflex manner. Consciousness provides a way to evaluate the world, to give attention to parts of it and focus the brain's activity and response, reducing multiple inputs and clues to a single perception. In short, conscious animals can make complex decisions and survive better. To demonstrate the point, a display in the Museum of Life on Earth shows a corvid, a scrub jay, burying food to eat later. This shows forward planning, but the bird also employs deception when hiding food to deceive rivals, showing it has an idea of self.

The octopus gives us a faint idea of the alien. It has come to represent something of the other, a strangeness that goes beyond a description of their anatomy and behaviour. Two-thirds of their neurons are in their eight arms which can operate semi-autonomously, as if there were eight conscious experiences associated with them partly unified by their brain. The octopus can change shape and colour, it tastes with its skin and has its mouth in its armpit. It is hard to imagine such an alien way of interacting with the world.

In the central spectacula, they all glide past – the panoply of life on Earth: a baby sperm whale pushes through a sargassum patch, eagles, tigers, squid, shrimps with their almost detached eyes and large antennae, insects and jellyfish, one of whose kind lays claim to being the only immortal creature on our planet. Predators and parasites, cellular and multicellular, creatures working in isolation and in exquisite cooperation, a stream of ants pass by all with a singular purpose. Creatures of the cold who hibernate to survive and those who live for just a day in tropical rainforests. Habitats with high life density contrasted with the sparseness of deserts. For any biologist, human or otherwise, our planet is a treasure, a lifetime's study whatever its duration.

Then come the most intelligent species. We share a level of intelligence with dolphins and orcas, whose last common ancestor with humans lived 85 million years ago. There are cephalopods with their large brains and remarkable learning skills.

Away from the central hall is a region devoted to the earliest form of life that appeared not long after the Earth and its oceans were formed. Many people imagine the early Earth as some version of hell (it is after all referred to as the Hadean era – from

Hades), with landscapes of lava oozing out of volcanoes and bolts of lightning punctuating a noxious atmosphere. The Earth was certainly like that for a while but it's difficult for a planet to stay in such an extreme state; it wants to cool and does so quickly within just a few million years. Scientists have studied tiny crystals called zircons which are splinters of the first rocks to solidify on Earth. They contain a snapshot of conditions at the time and testify to moderate conditions and the presence of water. Thus, the stage for life was set.

Perhaps life had more than one genesis on Earth. Not long after it was formed, the primitive Earth was struck by another world about the size of present-day Mars. The result was that the surface of our planet was sterilised, the metallic cores of these two worlds merged and the debris from the impact formed the Moon. The Earth cooled again, and perhaps life arose once more. If it did then this twin genesis has obvious implications for the frequency of life in space.

Is life inevitable given time? It's an important question for those interested in life in space. Some imagine it came from a soup of chemicals out of which emerged the basic molecules of Earth life, producing amino acids and the single-stranded RNA (ribonucleic acid), which was capable of carrying information. Perhaps RNA began to spread in warm ponds and eventually there were molecules that became surrounded by a fatty bubble – the first cells. Fatty acids, fairly abundant molecules on the pre-biotic Earth, can naturally form tiny enclosures as they have one end that hates water and another end that likes it, so they self-organise into spheres. This kind of membrane is critical for life because it isolates and concentrates chemical reactions, while allowing new ingredients to enter and waste to leave. Later, two strands of RNA intertwined and became the more familiar DNA.

We classify life on Earth into three grand domains which are the highest taxonomic rank: archaea, bacteria and eukarya – we humans are part of the latter. This has been likened to a tree that has the primordial life form that developed into these three domains at its base. Bacteria and archaea probably make up 90 per cent of the biomass of the Earth and comprise a biological world that is barely explored. There are more bacteria in our guts than all the humans who have ever lived.

Cells are the basic biological, structural and functional unit of all known organisms. They emerged about 3.5 billion years ago. As microbiologist Lynn Margulis said: 'To go from a bacterium to people is less of a step than to go from a mixture of amino acids to a bacterium.' Cells are far more complicated than stars. You have 40 trillion of them in your body, 80 billion in your brain.

In the museum, you are now looking at a cyanobacteria that in reality is 50 microns long. They date back 3.5 billion years and their descendants are still found in many environments. They are one of Earth's most successful organisms and helped produce almost all the oxygen in the atmosphere. Around 2.4 billion years ago during the Proterozoic era, a time of mountain building as a supercontinent was being assembled, oxygen appeared in substantial quantities in the Earth's atmosphere for the first time. With it, the nature of life began to change because life had changed the Earth in a poetic relationship thanks to photosynthesis. The oxygen was toxic to most life forms but the increased energy it provided to some removed the blocks on life's potential. The Great Oxidation Event, as it is called, is a key event in the development of complex life. The oxygen in your bloodstream – an efficient way of delivering energy to cells – is there because of those ancient cyanobacteria. Oxygen poisoned many earlier species, but it led to the development of the eukaryote cell. Unlike bacteria, these cells have a complex, convoluted membrane surrounding them and a nucleus which contains their genetic material. They can also form many-celled organisms. Some believe that oxygen is essential for multicellular organisms, but all we can truly say is that it is just the way it happened here. Perhaps this early story also applies to other worlds? It might be that oxygen will accumulate in the atmosphere of most oceanic planets as they age, fuelling the development of more advanced life forms.

One wing of the museum asks what are life's limits? On Earth, life is almost everywhere you look even in regions where you would think it was impossible. Life on Earth has found many ways to exist, and so our visit to the Museum of Life on Earth could tell us a great deal about life elsewhere in the universe.

We are now visiting the isle of the extremophiles. We are descended from extremophiles. The creatures closest to the root of the tree of life on Earth lived in conditions most life on Earth today could not endure. They are found in every environment, ever masterful at metabolising carbon as well as nitrogen, sulphur, iron and phosphorus in these extreme conditions. The oldest common ancestor of all life on Earth was probably thermophilic – it was able to tolerate great heat. Its descendants are still to be found in places like the hot springs of Yellowstone National Park or the superheated water found in deep-sea vents.

Extremophiles challenge our assumptions about life. One of them can grow in water temperatures of 121°C. Some scientists believe that there may even be extremophiles living near deep-sea hydrothermal vents that could tolerate temperatures up to 200°C. At the other extreme, ice worms move through Alaskan glaciers and methane ice deposits in the Gulf of Mexico using natural antifreeze to protect themselves. *Deinococcus radiodurans* can withstand radiation thousands of times more intense than that which would kill humans – it has five back-up copies of its genes ready to be used and is protected by a tough outer layer of lipids that can survive the vacuum of space. Microbes like this might have clung to rocky debris that passed between the planets and moons of the solar system and even beyond.

Life has also adjusted to living in the highest freshwater lakes on Earth. In the Licancabur Caldera in the Atacama desert, bacteria have adapted to the high UV flux at that altitude that usually damages molecules. Life is found on the deepest seafloor of the Mariana Trench, 11 kilometres deep, where the water pressure is 1,100 times the air pressure at sea level. Single-celled organisms called foraminifera are found in these sediments where they have been feeding on sunken organic matter for hundreds of millions of years.

The museum display changes again and you are now many kilometres below the Earth's surface to examine a kind of life that blurs the boundaries between chemistry, physics, geology and biology. It seems that radioactivity can sustain life by being able to split water molecules into hydrogen, peroxides and radicals – all sources of energy that can be used for metabolism. This once overlooked process of radiolysis has opened up whole new vistas into what life could look like and how it might have emerged on a primitive Earth. This new metabolic landscape is almost unexplored. By various estimates, the inhabited subsurface may be twice the volume of the oceans. The display now morphs to show Mars as seen from orbit, zooming in on regions that might be able to use radiolysis to sustain microbial ecosystems akin to those on Earth where future missions should be targeted. If life on Earth can exist without the benefit of sunlight, then we must change the traditional understanding of there being a habitable zone around a star.

Behold now a tardigrade with its five body segments, four pairs of clawed legs and single gonad. They have a multilobed brain and robust digestive and nervous systems. Discovered in 1773, more than 100 species are known, but there may be hundreds more yet to be found. They are creatures that have their own phylum. They are small, about 200 microns across and can live in all climatic zones, from the arctic to the rainforest, and can survive in temperatures from –270°C to 150°C and pressures from a vacuum to 1,000 atmospheres. They can also survive 1,000 times the radiation that would kill a human. They are one of the Earth's oldest and most successful life forms. They also have a space programme, or rather they have been placed on board the International Space Station and exposed to outer space, they have also travelled on an Israeli spacecraft that crash-landed on the Moon. Tardigrades are not extremophiles in the strictest sense of the definition. They stay alive waiting for better conditions. Some bacteria, spores and cysts appear to be able to survive for a million years in suspended animation.

One of the domains of life, the archaea, are like bacteria in that they are mostly single-celled creatures whose cells contain no nucleus, but they have unique properties that separate them from bacteria and biochemically they have more in common with humans than bacteria. They are able to use more energy sources than we eukaryotes. They are essential for us as they exist symbiotically in our gut.

As you enter the next auditorium in the museum, a crucial event is being staged: the origin of eukaryotes. These are organisms, like us, whose cells have a nucleus. Our cells are like cities with complex internal structures and regions that generate energy, build proteins and transport them around the cell. Eukaryote cells can come together and specialise in forming complex creatures that reproduce sexually and thus evolve rapidly. Now you are being shown an archaeon, a species of archaea that is growing tentacles in what might be one of the handful of crucial events in the development of life on Earth, and probably for many other places in the universe as well.

There is a theory that eukaryotes descended from an archaeon that merged with another microbe and this is what you are seeing being re-enacted in this Museum of Life on Earth. The archaeon is reaching out and surrounding the adjacent bacteria with its protrusions, eventually engulfing its neighbour and using its abilities to increase its complexity and thus its potential. This took place at the very root of the tree of life and shows us that we humans are an amalgam of what came before.

No Museum of Life on Earth would be complete without the dinosaurs. They dominated Earth following the Great Dying of the Permian–Triassic period, some 250 million years ago. This was a planet-wide reshaping of Earth's life possibly as a result of a vast volcanic eruption or meteor impact, no one is sure. Within 10,000 years of the event, dinosaurs had moved in to occupy the now vacant ecological niches as the great supercontinent of Pangea started to break apart. Both young and old visitors feel a shiver as a Tyrannosaurus rex runs towards them. It was not until the dinosaurs were wiped out 65 million years ago that the first mammals grew larger than the size of a squirrel. Mammals had different abilities than dinosaurs: an increased use of sight rather than smell, grasping hands, complex social lives. The development of acute, stereoscopic vision was probably a spur for the growth of larger brains; in humans, processing vision occupies 30 per cent of the brain's capacity. Visitors are reminded that not all dinosaurs died out; Avian dinosaurs survived. Today we call them birds.

Mammals are a small part of Earth's biota. The roughly 5,400 mammal species are outnumbered by 8,200 reptiles, 10,000 birds and 29,000 fish, while the sum of all 52,600 vertebrate species is dwarfed by 390,000 species of plants and 1.2 million species of invertebrates, 950,000 of which are insects! (Three-hundred thousand of those are beetles, which led the biologist J.B.S. Haldane to comment that God has 'an inordinate fondness for beetles'.)

One of the corridors off the spectacula leads to perhaps the most magnificent area of the museum, featuring relatively new inhabitants of planet Earth – the first remains of flowering plants are only 125 million years old. After life split into archaea and bacteria, the eukaryotes themselves started to diversify with protists and fungi splitting from the line that would become animals and plants. With so many colours and shapes from their 64 orders, their 416 families and about 13,000 genera comprising over 300,000 known species, the flowering plants are the flagship organisms of planet Earth.

The display features Amborella, which is found in New Caledonia. It is the only member of its family, a sister group to other flowering plants. With its red fruits, it lies near the base of the flowering plants lineage, relying on insect pollination and the wind to procreate. Also appearing are water lilies rooted in soil, whose leaves and flowers emerge above water – the pale red and white of the Santa Cruz water lily, and the crimson lily from Canada. All more vivid than Claude Monet painted them.

Land plants have existed for at least 475 million years and are dominant over large swathes of the Earth, notably the taiga of the northern hemisphere. There are over 3 trillion trees in the world. Worshipped in the past, it was said that spirits lived in trees. The oldest trees are over 5,000 years old.

The group of vascular plants, ferns, reproduce via spores and have no seeds or flowers. They appear in the fossil record in the middle Devonian period, about 360 million years ago. Damp and humid environments are home to the hornworts, liverworts, mosses and green algae. Many organisms rely on green algae to conduct their photosynthesis for them. The algae were ingested into their cells a long time ago and those cells still contain a nucleomorph or vestigial nucleus.

Next you are introduced to the bioluminescent creatures, starting with fireflies skitting around and then sedentary glow-worms. Night falls on a coral reef, which under blue light returns the pink and green glow of creatures unseen, and then those of the deep ocean, which carry luminescent lures and whose pulsations carry an unknown message. You then hear the sounds of the ocean, the clicks of whales and a climbing screech that touches something melodic and ethereal, high squeaks, base notes, the song of a male humpback whale, which may last half an hour, and sweeps across ocean basins and has done for 50 million years. In the darkness of the ocean, relationships are conducted by sound.

Your walk through the museum emphasises how little we know about some of the most common life on this planet. Consider the middle depths of the oceans that were once regarded as vast, empty, sterile wastes with only a thin rain of debris falling from the lighted zone above. We now know that this was wrong and was the result of previous sampling techniques being inadequate to survey this region and its now-known populations of delicate jellies that are barely visible to human eyes.

As we have seen, there are many moons in our solar system that have oceans of liquid water underneath a skin of ice. Principal among them are Europa orbiting Jupiter and Enceladus circling Saturn. The water is kept above freezing because of heat released in the interior of these tiny worlds due to tidal stress that is induced by their orbiting giant planets. Here we have environments that have been stable for tens, perhaps hundreds, of millions of years, along with nutrients and energy. Could there be life shielded in these bodies? One theory for the origin of life on Earth holds that it began next to hydrothermal vents which are commonplace on terrestrial ocean floors. After life began in the ocean it later colonised the land. I wonder how many worlds there could be like Europa and Enceladus in other planetary systems that have developed life only for it to remain trapped in a cage of ice. We know from whales and dolphins that intelligence can develop in an aquatic environment, albeit not our kind of intelligence. But could technology? Could a water-based species build machines? Are the creatures and philosophies of under-ice or water worlds to remain isolated with just their chemical, sonar and luminous abilities to exchange forlorn messages? We are beginning to develop the machines we will send to Europa in the next decade to drill through its ice shield. When it breaks through the basal ice into that sealed and dark universe, I wonder if its microphone will pick up clicks and chirps, a fugue from unseen creatures. I wonder if I will ever get to hear the songs of Europa?

Told in this museum are the stories of billions of years, of accidents, and many restarts, of creatures gone for ever (which is most of all life that ever existed) and the tales of the few survivors. It gives us some idea about what we might find on other planets around other stars but let us not pretend it will provide the full picture. As we shall see, we cannot define life, we can only describe it and even then we can only describe it incompletely. On other worlds, things may have started like they did on our world but taken a different course, or they may have been different from the start, or not even tethered to a planet at all. The enterprise of life, its adaptability and resilience raise the status of the Earth to cosmic significance, and among those life forms reside modern humans who dominate their planet for now and will hopefully continue to exist as long as it is habitable. Yet according to the biologist Stephen Jay Gould, writing in 1989, we came very close, thousands and thousands of times, to erasure by the veering of history down another sensible channel. Replay the tape a million times, he said, and I doubt that anything like *Homo sapiens* would ever evolve again. We are truly something our planet has not seen before.

As you look at the museum map and plan your tour through Earth's life, you realise that we are not an endpoint of life any more than any other currently living species represents an endpoint, for they all have billions of years of evolution behind them. Each individual can trace its lineage back to its ancestors and to the species it evolved from and on past the archaeon to Earth's first living thing. But as life evolved and branched out, most of life's experiments hit a dead end. If you were able to put all the species that ever existed in a timeline then you would see that most are no longer here, meaning that we, and much of the life we see around us, are but the youngest leaves on the tree of life.

Although the variety of species is much greater now than it was early on, the vast majority are still the simplest forms. From one viewpoint there has been no succession of life forms on Earth. The 'primitive' bacteria are still with us; indeed they are much more prolific in the ecosystem than any of the 'higher' organisms. Some biologists have suggested that the Earth is still in the bacterial age with more complex organisms being ecologically unimportant. It has been calculated that the viruses in the seas outweigh the whales by at least twenty times. Whales themselves have been around for only about 30 million years or so – though there were reptiles nearly as big, ichthyosaurs and plesiosaurs, which lived from 140 million until 65 million years ago. Before 300 million years ago there was nothing nearly as big as an ichthyosaur. Five-hundred million years ago there was little animal life on land, some plants but no trees.

There is a section in the museum about soils, an often overlooked but vital component of Earth's biome. They are a specialised ecosystem for degrading woody plant material, which back in time allowed the diversification of land plants, with fungi forming symbiotic relationships with mycorrhizae on tree roots to help them absorb minerals. The recycling of plant material was one of the most important steps in the evolution of life.

Now a hologram forms above you of a tiny, worm-like creature – the first ancestor on the family tree that contains most familiar animals today, including ourselves. *Ikaria wariootia*, 555 million years old and only 6 millimetres long, is the earliest bilaterian, or organism with a front and back, two symmetrical sides and openings at either end connected by a gut. Before *wariootia* there were mostly radially symmetric larger creatures, but this organism was a cellular experiment, and the invention of the anus was a great advance – a pass-through gut instead of a simple sac. This evolution had immense ecological consequences because the accumulation of faeces on shallow sea floors promoted a new ecology of filter feeders, along with their associated predators. New forms of life became possible because of the invention of the anus. Subsequently, in only a few tens of millions of years there occurred the origins of all the major kinds of cellular animals we have now.

Until the 1950s, most scientists believed that complex life began around 540 million years ago in what is called the Cambrian explosion. This was because there were almost no earlier fossils that had been discovered and almost all of the basic categories of animals alive today can be found in the fossils of 540 and 490 million years ago. However, newer research shows that something significant happened earlier that could have set the evolution of life running on a different course. The reason it didn't is one of the lessons of the universe and something we should pay attention to as we consider life in space.

The period between 635 and 540 million years ago is known as the Ediacaran period (named after the Australian hills where its fossils were first discovered), and it has become one of the most fascinating periods in the evolution of life on Earth. Ediacaran animals do not seem to have shells, which is one reason why their fossils are so rare. What's more, their structure appears different from animals from the Cambrian period. About half the creatures have anatomies that were never seen again, their variety extended the range of body plans beyond anything we have today. These alien designs show that different forms and different ways of life were possible on Earth and that we have today a fraction of what Earth's biology has been. These creatures once seemed to point to one possible future for life on Earth, but why were they suddenly wiped out? Perhaps they initially had no predators and never learned to fight so they were easy prey for subsequent predators. Their peaceful world had no future – a lesson for us as we consider life in space.

The ways of life portrayed in the Museum of Life on Earth are many and varied and obviously pay no regard for human prejudices. Xenopus is a frog that uses its own tadpoles as a food source, on the other hand, Alaskan salmon die after mating and laying eggs and the smell from their decaying carcasses prompts midges to lay eggs on the water surface with the resulting bacterial bloom serving as food for the hatching salmon.

Humans are not the only creatures represented in the museum's dome of intelligence. Studying dolphins alters any assumption that we are the sole proprietors of intelligence on Earth.

Humans have large brains, but certainly not the largest on the planet. Whales and elephants have brains up to six times larger, but they also have bigger bodies. Maybe the secret to greater intelligence isn't just a large brain, but the density of neurons within it. More penetrative studies have examined the number of neurons in the brains of different species and the amount of connections between those neurons that enable processing power. A human brain weighs about 1,300 grams and contains about 90 million neurons. Curiously, our brains have shrunk over the past 30,000 years, in which time our species has experienced its greatest development.

Dolphins separated from the evolutionary line of land-based mammals about 60–70 million years ago, when their ancestors, small, wolf-like animals, crept back into the sea. They actually had big brains for longer than humans. They understand symbolism, gestures that might mean the red or blue ball, and syntax, the idea that changing the order of words can alter their meaning. They appreciate their own individuality and have names that run in families. Studying them is the nearest thing we have to contact with an alien intelligence.

The next display takes you from the world of animals into something completely human in a way that is surprising to many visitors. What you see is not a life form but something closely associated with one particular form of life. It's a rainbow. Rainbows are not made of just sunlight refracted and reflected by raindrops acting as tiny prisms. What nature does is to fan out sunlight into its component wavelengths. The droplets do not produce colour. That is made inside our brains, in a particularly small part of the human brain that overlays the world with colour. Long before we were human, we lived in the forest canopy and only occasionally came down to the ground. Our world was one of green shapes and dappled sunlight. Our high-quality colour vision may have evolved in our wide-eyed primate ancestors as an aid to finding fruits in the predominantly green environment and also to see camouflaged predators as subtle changes in the tints of background colours. We were then, much more so than most of us are today, an integrated part of the ecology. We evolved to rely on the behaviour of some animals and the properties of some plants, and they in turn came to depend on us. The rainbow is a result of such evolutionary interdependence.

Consider rainbows from the perspective of plants. Of course, they cannot see a rainbow in the late afternoon arched over dripping, verdant tropical forests, or hung over the plains of the Serengeti after a passing shower. Yet somehow, they know it is there because it is as much their creation as it is of the sunbeam and the raindrop. Our improved colour vision was in a way driven by fruiting trees and bushes. Evolution used it to propagate the plants so that we would see, pluck and eat their fruits and scatter their seeds. Thus, the rainbow has been with us as long as we have been human. Much that is arbitrary and wonderful, parochial to Earth and yet universal, has come together to make rainbows, yet they are not here by accident. In a way, it is the true banner of humanity. No other creature in the universe sees a rainbow the way humans do. Other intelligences out in space will see their own rainbows but they will be theirs, they cannot experience the human rainbow. It is ours alone, the true flag of planet Earth and humanity, and it flies above the Museum of Life on Earth.

As you consider taking your leave of the museum, the display changes one last time and you see lazy waves lapping on white sand at the edge of a blue sea somewhere in the tropics 360 million years ago. It's obvious that this is a very different Earth from today. Forces operating from deep within the planet's interior resulted in all the land becoming concentrated in one region, as the dispersed continental masses of earlier geological epochs reunited and became bound together by great mountain ranges. Intense volcanic activity marked the closure of the once great Palaeozoic seas as the ancient continents of Laurentia and Baltica collided. For the first time, forests grew in the Earth's equatorial regions, and in the oceans it was the age of giant fish.

The Sun was slightly dimmer back then, the day shorter, the Moon closer to the Earth and the sky, had there been anyone around to judge, a shade bluer. It was a period of natural global warming, and everywhere was balmy and plentiful. Life of all kinds thrived in the warm seas. But the peace of this shoreline concealed a time of great change. The first animals to colonise the land were those that could not easily dry out, such as those encased in bony shells like sea scorpions. Free from predators and the hazards of the water, they grew up to 2 metres in size – fearsome unopposed creatures strode the semi-barren lands. Somehow, it seemed that the planet was experimenting with life, and in a sense, waiting for something.

The night sky was also different. There were more bright stars because the Sun was in a more crowded part of the galaxy than it is today. It was traversing the rich starfields of a galactic spiral arm. At night some of the stars were partially obscured by streams of gas that traced out coloured lagoons of light criss-crossed by black tangled tracts of interstellar dust.

Life had already won control of Earth after a struggle, the outcome of which was inevitable, already certain as soon as the primitive Earth had begun cooling. It is the nature of life to change and its ability to adapt is as constant as the stars, the galaxies and the vast darkness in between.

One day, a creature from the sea appeared at the edge of the water, its eyes poking above the waves, shifting from side to side taking in the strange nature of this new place. Soon the sunlight on its membranous skin was harsh and uncomfortable so it dipped back beneath the soothing waves and moved offshore to find its pack. The tide, more frequent and powerful than those today, quickly washed away its tiny footprints. In the few moments that its head was above the water, something stirred within the creature, some impulse it was unaware of but that was within it and also beyond it. It was an impulse that had not yet been seen on planet Earth, but yet had been in existence since perhaps before the universe was born. Will there always be new land, new territories, new planets to explore in the unoccupied and the beckoning expanses of the galaxy?

This creature that reached the land had already begun to change. It wasn't the case that evolution improved its fins, turning them into legs. The fins had already begun to be walking limbs while the creature was still using its gills in the sea. This is a process called 'pre-adaptation'. As a result of this, unexpected evolutionary pathways become available allowing further changes. In fact, most features originally evolved to carry out different functions from those they eventually performed. Those early limbs, with eight (or more) digits, evolved into the familiar five fingers, knees and elbows we have today, but it could have been different. The same goes for the 'face' with the nose above the mouth. Curiously, our imagining of aliens often feature knees and elbows and recognisable faces indicating that their origin would have been like ours.

They say that evolution involves long periods of time, but they are wrong. Its chief requirement is death, not time. In an instant, those eyes became something more. Sea creatures had looked at the land before, but this time, the creature that came back had a look that had never been seen on the Earth before. Someday humans would call it desire.

When these small events were happening in the tropics of Earth during the Devonian geological epoch, the galaxy was already old. Countless ages and stars had been born, gone through their life cycles and perished in bangs and whimpers and floods of radiation and matter. Planets had been raised in their midst, growing from rubble piles – some formed into worlds of rock and ice and some into worlds of gas before they were either cast into space or destroyed by a dying star. Life had arisen countless times and, in most cases, perished without record or trace. But, perhaps by some arbitrary law of chance and luck, some life had developed intelligence and some eventually became star-faring. And of those, a few star-hopped across the Milky Way in a time much less than the age of the galaxy. As we shall see, in the galaxy's time frame, life could have blossomed in one part of it and then, in a flash, be almost everywhere.

Perhaps the single stars, doubles and triplets, gas clouds, star clusters and galactic spiral arms all saw space probes of various origins and designs silently comb them, stopping off at star after star to carry out a survey or to replicate another probe that would be dispatched on its own mission to continue the quest begun millions of years before under the light of a now forgotten and increasingly distant mother star, and perhaps an extinct race.

Was there, at that time on Earth, orbiting high above this primitive planet, something that watched the shorelines of the newly assembled continents of the Devonian age? Did alien sensors record those first steps in the conquest of the land? Do those records still exist in some vast and ancient database held on some distant planet, or stored in some seldom-visited archive held along silent corridors of light in a vast structure plying between the scattered stars?

And, I wonder, is that data ever studied by alien scholars? Or is the conquest of the land on a non-descript planet of such little consequence, having been repeated so many times on so many worlds, that it was merely catalogued, cross-referenced and ignored? Perhaps if humanity, or what it is to become in the almost unending millennia ahead, survives into a distant future to make contact with other civilisations, it will find those records – a survey of the Earth in the late Devonian geological era – and look for the first time with meta-human eyes at the first land creatures swarming around tidal pools, marooned by the ebb tide. Their ancestors.

After much effort, with its skin cracking under the harsh Sun and with its gills aching for the comfort of the sea, the creature dragged itself onto the sand and fell, panting as the alien dryness scoured its skin and reached for its gills. Perhaps it died there. Perhaps the others who followed it also died.

But if you delve deeper into the alien archive, expand its timeframe and look at the highest resolution images, you will see that, eventually, a few of these creatures survived and colonised the land. Their fins and flippers became legs and arms and their lineage became a new line of evolution that would one day alter the face of the planet, producing creatures that would travel to the stars to meet others with whom they shared a common birth, and others with whom they didn't.

That meeting – the first contact between aliens – is not without risk. Contact between two perspectives of evolution begun under the light of different stars would seldom be easy, or equal, and would never be without danger. First contact is a risk most species go through. We will face it, perhaps not for a million years, perhaps tomorrow.

None will emerge unchanged.

Not all will survive.

THE COSMIC CAVE

'We scan the time scale and the mechanisms of life itself for portents and signs of the invisible. As the only thinking mammals on the planet – perhaps the only thinking animals in the entire sidereal universe – the burden of consciousness has grown heavy upon us. We watch the stars, but the signs are uncertain. We uncover the bones of the past and seek for our origins. There is a path there, but it appears to wander. The vagaries of the road may have a meaning, however; it is thus we torture ourselves.'

– LOREN EISELEY, 1946

For a million and a half years the Danube was a mighty river. During the ice ages, it began bringing down an increasing amount of debris that started to fill in the valley it had created. With the uplift of the Swabian Jura mountains, it started to use a new route some distance to the south. Consequently, today the Blau, Schmiech and Ach rivers use a valley which they never could have excavated. Modern humans used the Upper Danube as an entry corridor into Europe from the Middle East during a temperate period just before an extremely cold climatic phase about 40,000 years ago. They did not find the land unoccupied.

Perhaps the humans that were already there thought that when the others came they looked a lot like themselves, but not quite. They did not know them. They were different in ways they did not quite understand. They copied some things from them and the newcomers learned some things in return, but nothing made them like the newcomers. They had things the new arrivals did not. It was agreed that it was better before they arrived. They had survived the cold times, learned the hard way, but the newcomers were a different puzzling change. They had no words to comprehend them. They had a strangeness they had never seen before. In a way, the new ones were like a question answered differently.

The recently extinct species of human – Neanderthals – lived in what we call the 'remembered present' in north-western Europe up to about 35,000 years ago. They lived in small groups and had a primitive language and a culture more developed than many have given them credit for. They were not brute savages and slowly, so slowly, their society was changing, but not fast or far enough to cope with new kinds of change. They had been here for almost 200,000 years and had become a part of the local habitat. Little altered from brief life to brief life. Nothing really changed except family and few lived very long. Most suffered brutal deaths by predators or injury. They ritually buried their dead, but they did not think much of the past or the future. Their stories were primitive, and they could not organise themselves into large groups. They preferred family and village. They had their customs, but longer-term planning was beyond them. They were here and now, in the landscape and in their minds – the remembered present. There must have been contact between these two species of human, but despite a major effort to identify archaeological evidence of interaction between Neanderthals and *Homo sapiens*, we have yet to find many examples outside of the mingling of DNA.

The new humans brought with them a new kind of society with cosmology and religion, and a sense of superiority.

They were neither aliens nor invaders, but they might well have been. The Swabian caves they lived in have produced the earliest record of figurative art and music, a cultural blossoming at around the time of the extinction of the Neanderthals. Perhaps that was an inevitable result of a small demographic imbalance that wiped them out in about 30 generations or about a thousand years. Neglect, competition and inter-breeding played a role, as did genocide, as some surmise. Whatever really happened, the relationship between Neanderthals and *Homo sapiens* was different from anything we can experience in the world today. It perhaps provides a lesson for us for alien encounters with a more advanced species.

There is one artefact that tells you all you need to know about modern humans, their fresh capabilities and indeed their future. It is a sliver of carved mammoth tusk discovered in 1979 near an ancient fireplace in a cave under a layer of bone ash and close to tools and other artwork. The hearth was later sealed by a ceiling collapse. There are notches on all its edges suggesting that it is complete and not a part of a larger object. It looks like a man with a lion's head with arms and legs outstretched – an imagined creature. Some have called it the Adorant. It could be the oldest star chart we know of, at over 33,000 years old. It looks like the constellation we now call Orion the hunter. On one side, there remain tiny spots of manganese and ochre and the grouping of notches suggests a time-related sequence. There are 88 notches – the number of days in three lunations and the number of days Orion's prominent star Betelgeuse disappears from view each year. Orion was visible for a nine-month period each year, which indicates that this could be used as a pregnancy calendar to ensure that a birth took place after a severe winter in time to build up reserves during the better months. The nature of the object's weathering suggests the piece may have been carried in a bag, something precious not to be left around; perhaps a tangible reminder of the conjoined cycles of cosmic power and human fertility.

It was a long time ago when the first hominid looked up at the night sky with curiosity, but around 50,000 years ago, something happened to humans, a blossoming of culture and capabilities, a new outlook, a new society. Stone tools started to become more specialised into projectile points and blades. By 40,000 years ago the change was unstoppable, the appearance of cave paintings, figurines, musical instruments, all the product of a new-found imagination. We had become curious, super-curious. It has become something of a cliché that in almost every film about the dawning of human consciousness there are a group of early humans huddled around a fire watching the sparks drift upwards directing their gaze to the stars beyond, to those other campfires in the sky. I don't think they needed anything to pull their gaze skywards at night.

Some believe that our modern mind was forged partly by our relationship with the night sky and that connection began in caves – the Cosmic Cave if you like. The stars and our attempts to extract meaning from them forced forward our thinking and curiosity. Consider the development of hand axes. They changed little for almost a million years, showing that our ancestors remained intellectually dormant, never experimenting or wanting more than they had. But about 40,000 years ago, in the darkness, by the light of flickering torches in caves, and in a more organised way of hunting, something kindled human creativity. Our brain's pre-frontal cortex was growing to accommodate the new connections we were making. Along came symbolic expression and the idea of the unworldly, the abstract. As soon as humans painted the stars on the walls of a cave our journey to them was inevitable.

Caves were not just places of safety for ancient humans. They were also their interface with their gods and the universe. Their walls were the edges of another existence. Caves were also a representation of the female reproductive tract with their tight entrances and mysterious internal chambers. The cloudy water dripping on the walls was associated with fertility. Caves were ancient planetaria, observatories from which to explore the growing universe of our minds. Look at the famous caves of Lascaux. On the walls, a shaman has represented himself as a birdman, acquiring head and claws and stretching his arms like wings to indicate a transformation depicted as ithyphallic because he produces the life-giving substance of the Milky Way, thus helping to protect and renew cosmic fertility. Much later, the Ancient Greeks saw the Milky Way as milk spilled from the breast of the goddess Hera.

The desire was to come into contact with some primeval power, something alien and the prototypes of all creatures. It drove shamans into the power caves – a term used by the people of southern-central California. Their belief was that the inner forces of nature and creatures of other worlds existed in caves. They wanted to contact these alien intelligences, these totem ancestors, and to read the mind maps of the world, which they thought were written on the walls. They considered the depths of the cave to be at the very centre, heart and origin of the world, the womb of the universe. We have always wanted to believe in aliens.

We have worked out many things on the walls of these caves. Anyone who has been faced by their spectacular beauty could be forgiven for not noticing many small symbols and signs scattered among the drawings of horses and bison. Of late, they have garnered some attention along with the thinking that they might be significant. Strangely, some 26 signs have been found in caves over a wide geographical area in southern Europe, originating from 35,000 to 10,000 years ago, which is coincident with the arrival of modern humans. Straight lines, circles, triangles, wavy lines and more complex signs could be telling us something we have overlooked. Perhaps they mark a dawning realisation that things can be represented symbolically as well as in literal drawings of animals and that symbols might represent something of another world yet connected to ours. Eighteen of the symbols have been detected in Australia from 40,000 years ago, which perhaps points to a creative explosion that took place independently in various parts of the world.

These symbols were part of a process that enabled people to share information beyond an individual's lifespan – special symbols, revered and sacred, telling special stories across their short generations. We will probably never know what they mean. I wonder if there are also symbols passing in between the stars and, if we found them, would we ever be able to determine what they meant?

Thus, the first human cosmic connection was forged, and it has influenced the way we see the universe ever since. But we are still a species in transition. When our minds changed 50,000 years ago, we learned to face the universe and ask questions, solve its problems with an absence of emotion and using an objectivity combined with linear thought and sustained attention spans. We made our own consciousness, but how much of it is just human and how much has a universal component that we would share with any alien we might find? How much do we share with the creatures of our own planet? Is there a spectrum of consciousness, and, if so, where are we on it? William James, the father of American psychology, said: 'No account of the universe in its totality can be final which leaves these other forms of consciousness quite disregarded.' Perhaps the most interesting question concerns not how aliens evolved but what their mentality, their consciousness, is like. What realms of consciousness could be generated by other nervous systems? As to the mind of an alien, nobody knows anything. Or perhaps we do, perhaps we have to search ourselves as well as the stars for an answer. Hamlet argues with himself, not the gods.

In Plato's book *The Republic*, he relates how Socrates described a group of people who lived their lives chained to a blank wall. They watched shadows projected on the wall from objects passing in front of a fire behind them and they gave names to these shadows. The shadows are their reality, but the shadows were not the real thing. Would aliens be like the newcomers to the Neanderthals in north-western Europe or like shadows on the wall of a cosmic cave?

SPEAK TO ME

Alien contact could come in many guises. Instantaneously or gradually. Perhaps it will be something so obviously artificial and unhuman that it immediately shows itself to be the mark of an alien intelligence – a signal or an artefact of some sort. Perhaps it will be a dawning realisation that something already in plain sight can be interpreted differently, revealing a previously unappreciated design or interference. However it comes about, in that moment of realisation our conception of the universe would change. It is the thing that scientists, artists and philosophers crave. One person in all of history would make that first contact. Life would have a new story to tell, new doors and possibilities would open to wonders and fears. All that happened before it would be part of a different universe, a sterile one in which we were alone yet supreme. Afterwards, as far as our evolution goes, it would be year zero.

As we shall see, we have been blinkered in the way we might discover aliens as we have placed too much importance on one potential method of communication. The way most favoured by scientists is that we would detect an artificial radio signal coming from space. Some are surprised we have not yet detected such a signal, others are glad we haven't. On more than one occasion we thought we had detected such a message and felt the thrill of first contact.

In 1901, the eccentric genius Nikola Tesla believed he had received a message from aliens by way of newly discovered radio waves. As he put it: 'It was sometime afterward when the thought flashed upon my mind that the disturbances I had observed might be due to intelligent control … The feeling is constantly growing on me that I had been the first to hear the greeting of one planet to another … "Brethren! We have a message from another world, unknown and remote. It reads: one … two … three …" The following year, the distinguished British physicist Lord Kelvin was in the United States and announced that he believed Tesla had detected Martians calling Earth. He said that New York was the 'most marvellously lighted city in the world' and that it could be seen from Mars. 'Mars is signalling … to New York,' he said.

Two years before his purported alien signals, Tesla had set up a laboratory in Colorado Springs equipped with a 200-foot transmission tower with high-voltage equipment to generate radio waves. He wanted to electrify the Earth so that anyone could place a conductor into the ground and tap our planet's electrical potential. Free energy for all was his wish, but Tesla was naïve. Although he could master the strange powers of magnetism and electricity, the human forces of money and monopoly, greed and power were a mystery to him. One night during his research, when he was alone among the things he did understand, his emotionless coils and transformers, he observed 'electrical actions':

> The changes I noted were taking place periodically, and with such a clear suggestion of number and order that they were not traceable to any cause then known to me. I was familiar, of course, with such electrical disturbances

> as are produced by the Sun, Aurora Borealis and earth currents, and I was as sure as I could be of any fact that these variations were due to none of these causes.

Later, *The Richmond Times* reported:

> As he sat beside his instrument on the hillside in Colorado, in the deep silence of that austere, inspiring region, where you plant your feet in gold and your head brushes the constellations – as he sat there one evening, alone, his attention, exquisitely alive at that juncture, was arrested by a faint sound from the receiver – three fairy taps, one after the other, at a fixed interval. What man who has ever lived on this earth would not envy Tesla that moment!

The question of whether there is intelligent life in space reached a turning point during the last decade of the 19th century as advances in telescope technology allowed more detailed scrutiny of the planet Mars. One of the chief protagonists for life on Mars at the time was Bostonian astronomer Percival Lowell, who later wrote that life waits in the wings of existence for its cue to enter the scene the moment the stage is set. And the stage had been set.

Giovanni Virginio Schiaparelli was the greatest observer of Mars in the 19th century and during the favourable close approach in September 1877, when the planet was just 56 million kilometres away, he produced a remarkable map. Others had seen hazy patches and little detail on the disc – but

Schiaparelli saw more. He saw lines that he called *canali*. The Italian word means 'channels', and does not imply an artificial origin, but others translated the term as 'canals', and soon speculation was rife about a dying planet whose inhabitants constructed giant canals to bring water from the polar caps to the equatorial deserts. Also watching Mars that year was Lowell, who became obsessed by the canals, a little too obsessed it turned out. When he heard that Schiaparelli had been forced to discontinue observing due to failing eyesight, he decided to build an observatory to study Mars. It seemed obvious to him and to many others that Mars was another kind of Earth. The two planets had so much in common. Mars had an atmosphere and Earth-like markings on its surface, including what seemed to be polar ice caps. Its day was only 41 minutes longer than ours.

The 1892 opposition was the most favourable since that of 1877 and Lowell's assistant Andrew Douglass travelled to Peru determined to see the canals. He saw even more spots at the intersection of canals as well as lakes. In 1893, the Irish astronomer and prolific popular lecturer Sir Robert Ball wrote: 'That there may be types of life of some kind or other on Mars is, I should think, very likely … speculations have also been made as to the possibility of there being intelligent inhabitants on this planet, and I do not see how anyone can deny the possibility, at all events, of such a notion.'

At Lowell's new observatory in Flagstaff, Arizona, Mars was observed almost every night between June and December 1894. Lowell had driven the construction of the new observatory on a high plateau in northern Arizona at record speed. He and

Douglass produced over 900 drawings confirming his belief that he was seeing a planet-wide irrigation network. 'Without seas and mountains, life would tend the quicker to reach a highly organised stage. Thus, Martian conditions make for intelligence,' he wrote. The public were convinced there was life on Mars thanks to Lowell's popular book *Mars*, released in 1895. In his 1897 novel *On Two Planets*, Kurd Lasswitz imagined the Martians would help humans reach a higher state. H.G. Wells, on the other hand, in his 1898 novel *The War of the Worlds*, foresaw conflict and disaster. This division of opinion about the outcome of extraterrestrial contact has continued. But by the time of the next Mars approach, others could not confirm Lowell's findings. Alfred Russel Wallace wrote: 'Mars, therefore, is not only uninhabited by intelligent beings as Mr Lowell postulates, it is absolutely uninhabitable.'

Likewise, Tesla's alien signals captured media attention but did not impress scientists, despite the fact he was not alone in making such claims. Another pioneer of radio, Guglielmo Marconi also said he had contacted Martians. Thomas Edison later said that Marconi's work was 'good grounds for the theory that inhabitants of other planets are trying to signal to us'. Tesla welcomed contact with aliens: 'Faint and uncertain though they were they have given me a deep conviction and foreknowledge, that ere long all human beings on this globe, as one, will turn [their] eyes to the firmament above, with feelings of love and reverence, thrilled by the glad news.'

In 1901, Marconi and his radio waves had spanned the Atlantic and a revolution in communications was underway, so when the First World War arrived, transatlantic radio signalling was routine. In 1920, Marconi said his stations on both sides of the Atlantic had been detecting strange signals since before the war. Some were in what sounded like a code but were meaningless as far as he could tell. When asked by a reporter if these might be from another planet, Marconi said yes, adding: 'If there are any human beings on Mars I would not be surprised if they should find a means of communication with this planet.' On 2 September 1921, *The New York Times* reported that Marconi was certain the signals came from Mars. Tesla was unimpressed with Marconi's assessment, saying they could be 'undertones' of man-made signals, like harmonics of a musical note. It all added to the excitement of the time. There was to be another close opposition of Mars in 1924, and by this time we had radio waves. It was the best chance humankind had ever had to contact Martians.

In the 17th century, Voltaire called Bernard Le Bovier de Fontenelle 'the most universal genius that the age of Louis XIV has produced'. Fontenelle helped write an opera at the age of twenty and later became official historian of the French Academy of Sciences. The work that brought him fame throughout Europe was called *Conversations on the Plurality of Worlds*, published in 1686 as a series of discussions with a fictitious marchioness. He maintained that the worlds of our solar system were inhabited. Those who lived on Mercury – the nearest planet to the Sun – 'are so full of fire, that they are absolutely mad: I fancy they have not any memory at all', he stated. He wrote that the residents of Saturn, outermost of the known planets, 'live very miserably. The Sun seems to them but a little pale star, whose light and heat cannot but be very weak at so great a distance.' He also said that visitors from other worlds travelled by comet. 'You know all is very well', the marchioness replied, 'without knowing how it is so; which is a great deal of ignorance, founded upon a very little knowledge.' Even in the Enlightenment, such ideas were still dangerous. It was, after all, less than a century since Giordano Bruno had died at the stake for heresy.

Swedish theologian and mystic Emanuel Swedenborg had a somewhat different view. In a series of dreams that he regarded as divine revelations, he said he was visited by spirits from other planets who told him of countless inhabited worlds and that there were two races living on Venus: one gentle and humane; the other savage and cruel. Those inhabiting Mars were the finest residents of the solar system, resembling the early Christians in their piety.

On 25 August 1835, there was an article in the daily *New York Sun* with the headline 'Great Astronomical Discoveries Lately Made by Sir John Herschel at the Cape of Good Hope'. It was written by Richard Adams Locke. The newspaper reported that the English astronomer had, by using a 7,000-kilogram telescope lens with a magnification of 42,000, been able to observe rocks, trees, flowers and even intelligent winged beings of both sexes on the Moon.

New York went into a frenzy at the news – after having read only one instalment. Locke could hardly believe his luck. Building on the excitement, he added that the *Sun* had an exclusive series of reports about Herschel's latest discoveries. Excited readers certainly got their money's worth with accounts about one spectacular lunar discovery after another. The second instalment revealed that the Moon was covered with crimson flowers 'precisely similar', remarked Herschel's assistant Dr Grant, 'to the Papaver rhoeas, or rose-poppy of our sublunary cornfield'. Poppies were not all that the Moon had to offer, indeed, there was an entire lunar forest. 'The trees,' said Dr Grant, 'for a period of ten minutes, were of one unvaried kind, and unlike any I have seen, except the largest class of yews in the English church-yards, which they in some respects resemble.' The observers also saw a green plain on which some kind of fir tree grew and then a lake of marine-blue water. In the next valley, Herschel and Grant saw groups of crystals, and amethysts rose from the ground in the shape of obelisks and pyramids. In the distance were herds of tiny bison, their shaggy pelts covering their faces to protect their eyes from the extremes of light and darkness on the Moon. Then blue goat-like creatures appeared. The second article in the *Sun* ended with a description of the unicorn that lived on the Moon. New Yorkers loved it. The circulation of the *Sun* climbed from 8,000 to 19,360; it had become the newspaper with the world's largest circulation overnight. Even *The Times* of London had a circulation of only 17,000. And Richard Adams Locke had barely begun. The next article reported on the discovery of intelligent life:

> We were thrilled with astonishment to perceive four successive flocks of large winged creatures … descend with a slow even motion from the cliffs on the western side, and alight upon the plain … We counted three parties of these creatures … walking erect towards a small wood near the base of the eastern precipices. Certainly they were like human beings, for their wings

> had now disappeared, and their attitude in walking was both erect and dignified ... These creatures were evidently engaged in conversation; their gesticulation, more particularly the varied action of their hands and arms, appeared impassioned and emphatic. We hence inferred that they were rational beings.

Locke told them of the Temple of the Moon, constructed of sapphire, with a roof of yellow, resembling gold. There were pillars 21 metres high and 1.8 metres thick. More 'man bats' were seen walking through its precincts. But by now even Locke could not continue with his fantasy. Readers of the *Sun* who were eagerly awaiting more astounding detail were told that the telescope had, unfortunately, been left facing the east and the Sun's rays, concentrated through the lenses, burned a hole '15 feet in circumference' through the reflecting chamber, putting the observatory out of commission.

Later, Edgar Allan Poe explained that he stopped work on the second part of his *The Unparalleled Adventure of One Hans Pfaall* because he felt he had been outdone, and one Harriet Martineau of a Springfield, Massachusetts missionary society resolved to send emissaries to the Moon to convert and civilise the bat men. Then the *Journal of Commerce*, another New York paper, wanted to reprint the entire story. At first Locke tried to dissuade the editors from running it – it was yesterday's news, he argued. They persisted and eventually Locke was forced to tell the truth. The *Journal of Commerce* got its scoop by announcing that the whole story had been a fabrication, from beginning to end. For a while, many believed there was intelligent life on the Moon and it was met with fascination, not panic.

But how do we signal across the gulf of space? If there are bat men or Martians, how do we let them know we are here? The mathematician Carl Friedrich Gauss proposed that forests be modified in Siberia to make a gigantic right-angled triangle that could be seen from Mars. In Vienna, the astronomer Joseph Johann von Littrow had a similar notion that canals be dug in the Sahara, forming geometric shapes. In France, Charles Cros wanted to build a vast mirror to reflect sunlight toward Mars. A similar idea was suggested by the Sperry Gyroscope Company after the First World War, in which it would use its high-powered searchlights to direct a beam towards Mars.

The close opposition of Mars in 1924 approached. In anticipation, David Todd, who had been head of the astronomy department at Amherst College in Massachusetts, proposed all radio stations on Earth be shut down to listen for signals. On 21 August 1924, the chief of naval operations of the US Navy sent a message to the most powerful stations under his command, from Cavite in the Philippines to Alaska, from the Panama Canal Zone to Puerto Rico, telling them to avoid unnecessary transmissions and to listen for unusual signals. A similar order was sent to army stations.

Charles Francis Jenkins was waiting with his radio receiver. As a 29-year-old, he may have been aware of a 1896 newspaper article entitled 'A Signal from Mars', which reported on a 'luminous projection on the southern edge of the planet', suggesting 'the inhabitants of Mars were flashing messages' to Earth. We can find this same idea in a piece of music. A few years later, someone composed 'A Signal from Mars, March and Two Step'. Jenkins was one of the pioneers of television and held over 400 patents in the fledgling technology. In 1928, he established the first TV broadcasting station in the United States, W3XK, but his system relied on mechanical devices that were soon superseded by electronic components, consigning him to be one of television's lesser-known developers. But although his TV system did not communicate with many on Earth, he did try to establish communication with intelligent life on Mars. In 1918, he built a radio receiver, the extraordinary SE-950, which can still be seen in some museums. It was designed to be a rugged field radio for the US Navy in the First World War as it could also be used to locate enemy radios. It is a magnificent piece of work for its time with its window showing the glowing valves and its Bakelite dials and knobs. It had some unique features, such as different knob styles for different sensitivities as well as numerous external terminals for connecting to other devices. After the war, its versatility allowed it to remain useful in Jenkins' laboratory.

Between 21 to 23 August 1924, Mars came closer to Earth than at any time in the century before or for the next 80 years. In the United States, it was 'National Radio Silence Day' when for the 36 hours all radios went quiet for five minutes on the hour. At the US Naval Observatory in Washington, a radio receiver was lifted 3 kilometres above the ground on a balloon. William F. Friedman, the chief cryptographer of the US Army, was assigned to translate any potential Martian messages picked up by the Jenkins' SE-950, which was connected to a device capable of photographically recording any 'alien communication'.

Nothing was heard from Mars, but something had shifted. The technology of radio communication and its possibilities set in motion a train of technological developments that changed the way we looked at the universe.

Radio astronomy came about by accident. Karl G. Jansky, of the Bell Telephone Laboratories, was trying to track down the source of interference in the company's transatlantic radio communications. In 1931, at Holmdel in New Jersey, he built an antenna array on a wooden frame that was 100 feet long, which rode on four wheels taken from a Ford Model T. He called it his 'Merry-Go-Round' and it rotated every twenty minutes. He found three types of interference. One came from lightning in nearby thunderstorms, the second from more distant storms, but the other form was different. He put it on a loudspeaker and noticed that it slowly changed in intensity during the day, noting that it moved 'almost completely around the compass in twenty-four hours'. In January, he said, its direction coincided with that of the Sun, but a few months later it was coming from a different direction. He then realised that it remained fixed among the stars. One time there seemed to be a jump in his record until he realised that he had forgotten to allow for Daylight Saving Time.

He announced his discovery at a meeting of the American Section of the International Scientific Radio Union on 27 April 1933. The radio emissions, he said, appeared to be coming from beyond the solar system. They might be from a single source or 'from a great many sources scattered throughout the heavens', he explained. Jansky's report of signals from beyond the solar system was front-page news, although he discounted the possibility that they were of alien origin. Most astronomers were unimpressed. It was left to an amateur astronomer, Grote Reber, to take the next step. Reber was a radio engineer and built a parabolic antenna in his backyard at Wheaton, a Chicago suburb. It was the first dish that was turned on the heavens and his results, published in 1940, showed almost the entire Milky Way to be a source of radio 'noise', with several regions of more intense emission. In the Second World War, the Germans had been trying to defeat the British warning system by flooding its radar receivers with signals. It was noticed by the British that the direction of the jamming pointed to the Sun as well as the continent. This was a military secret.

In 1944, news of Reber's work reached the German-occupied Netherlands and the astronomers at the Leiden Observatory. Jan H. Oort, director of the observatory, held a seminar on the implications of Reber's observations. An entirely new window had been opened on the heavens, Oort said. Until then, all of our knowledge of the universe, apart from meteorites or cosmic-ray particles, had been acquired through observation of that narrow band of the electromagnetic spectrum called visible light. Oort knew a revolution was coming, but there was a problem. Unlike light, the radio spectrum seemed to lack the bright lines that were so useful in optical astronomy. Hendrik Christoffel van de Hulst then suggested that clouds of individual hydrogen atoms should naturally emit radio waves of a wavelength of 21 centimetres and since it was believed that hydrogen clouds are commonplace there should be a sharp increase in radio emission at that wavelength. By good fortune, this wavelength passes freely through space and the Earth's atmosphere.

To find the 21-centimetre line, a Harvard University carpenter built a horn antenna on the laboratory roof at a cost of $400, using borrowed equipment. On 25 March 1951, when it was turned to the sky, the predicted emissions were there. Less than two months later, the Netherlands astronomers obtained similar results and both groups found that the emission, instead of showing up as a single, narrow line at 21 centimetres, was spread out in wavelength, with several peaks of intensity. What they were seeing was the Doppler Effect on hydrogen clouds moving at different velocities towards and away from the receiving antennas. The relative motions were produced by the Earth's spin, by our planet's orbit around the Sun, by the Sun's orbit around the core of the galaxy and by the movement of the hydrogen clouds themselves. Taking all of this into account, the Dutchmen were soon mapping the spiral arms of our galaxy that are hidden from us by dust clouds and produced a magnificent map showing the unseen spiral structure of our galaxy. The discovery of the 21-centimetre line was a tremendous boost to radio astronomy.

In England, the giant radio telescope at Jodrell Bank – the brainchild of Sir Bernard Lovell, who had arrived in the muddy fields south of Manchester with trailers of Second World War surplus radio equipment intent on studying radar echoes from meteor showers – was already under construction when the news was received that 21-centimetre emissions had been observed. The dish was originally designed to reflect much longer radio waves and a wire mesh was sufficient to reflect them to the focus of the dish, but the design was changed to solid metal sheets that would reflect radio waves of centimetre wavelengths. In the United States, the only large dish capable of observing at that wavelength was on the roof of the US Naval Research Laboratory, across the Potomac River from Washington's National Airport. It was soon realised they needed a bigger radio telescope in a better location, so the National Radio Astronomy Observatory in Green Bank, West Virginia was born.

Two astronomers who were considering how to search for life in space now enter our story. They were Giuseppe Cocconi and Philip Morrison, both professors at Cornell University in Ithaca, New York. In the spring of 1959, Cocconi was working on a paper he was due to present at a conference in Moscow. He believed that gamma rays – electromagnetic radiation of much greater energy than light – should be detectable from the Crab Nebula, the wreckage of an exploded star. As he discussed his idea with his wife, Vanna, herself a physicist at Cornell, it occurred to them that the scarcity of gamma-ray sources in the sky made it a promising method for interstellar signalling by aliens. Such sources were rare in the cosmos so would be easy to spot. The Cocconis had often discussed with Morrison the possibility of intelligent life in space. Morrison had been at Los Alamos as part of the project to build the first atomic bomb. He was interested in the idea of extraterrestrial life but pointed out that because of their high energy gamma rays would be difficult to generate and detect. He suggested they look at radio waves. They were low energy, easy to produce and receive and could be seen over vast cosmic distances. They decided to write to Bernard Lovell at Jodrell Bank with their ideas. The letter was dated 29 June 1959:

Dear Dr Lovell,

My name is probably unknown to you, so let me start by saying that I am now at CERN [The European Organization for Nuclear Research] for one year, on leave from Cornell University, where I am professor of Physics. Some weeks ago, while discussing with colleagues at Cornell ... I realised that the Jodrell Bank radio telescope could be used for a program that could be serious enough to deserve your consideration, though at first sight it looks like science fiction.

It will be better if I itemise the arguments.

1) Life on planets seems not to be a very rare phenomenon. Out of ten solar planets one is full of life and Mars could have some. The solar system is not peculiar; other stars with similar characteristics are expected to have an equivalent number of planets. There is a good chance that, among the, say, 100 stars closest to the Sun, some have planets bearing life well advanced in evolution.

2) The chances are then good that in some of these planets animals exist evolved much farther than men. A civilisation only a few hundred years more advanced than ours would have technical possibilities by far greater than those available now to us.

3) Assume that an advanced civilisation exists in some of these planets, i.e., within some 10 light years from us. The problem is: how to establish a communication?

> As far as we know the only possibility seems to be the use of electromagnetic waves, which can cross the magnetised plasmas filling the interstellar spaces without being distorted. So I will assume that 'beings' on these planets are already sending toward the stars closest to them beams of electromagnetic waves modulated in a rational way, e.g. in trains corresponding to the prime numbers.

They wrote that if a radio dish like Jodrell Bank were used to transmit from a distant planet and a similar instrument was on the receiving end, the power required to transmit detectable signals would lie beyond the reach of present technology. 'But I want to have faith and will assume that they have larger mirrors and more powerful emitters and can do it.' They suggested 'a systematic survey of the stars closest to us and spectroscopically similar to the Sun, looking for man-made signals'. 'All this is most probably fiction,' they wrote in conclusion, 'but it would be most interesting if it were not ... I leave to you the judgment on the feasibility of such a search.'

Lovell sent a brief reply that Cocconi later said was 'rather disappointing'. He said it would be difficult to do, adding that the radio telescope was carrying out 'a survey of certain flare and magnetic stars'. When that was over 'perhaps we shall have time to look at a few others' he explained. Later, Lovell did warm to the idea, but he wrote that he 'would still find it difficult to justify the diversion of any of the world's present radio telescopes to such speculative work. Nevertheless, during the past two years or so the discussion of the general problem of the existence of extraterrestrial life appears to have become both respectable and important.'

The idea was out there: search for a radio signal, but at what frequency? This was the problem with the original idea. Cocconi and Morrison realised that the answer had been with them all along. They submitted a letter to the scientific journal *Nature*, which stated: 'Just in the most favoured radio region there lies a unique, objective standard of frequency, which must be known to every observer in the universe: the outstanding radio emission line 21 centimetres.' Their article was published in the issue of 19 September 1959 and brought the debate into the open. The press was fascinated.

The ideas formulated at the time remain those dominant even today. Cocconi and Morrison said that some civilisations could last 'for times very long compared to the time of human history, perhaps for times comparable with geological time', and that they could achieve levels beyond our imagination and perhaps they are waiting for the emergence of intelligent life in our solar system. 'We shall assume,' they wrote, 'that long ago they established a channel of communication that would one day become known to us, and that they look forward patiently to the answering signals from the sun which would make known to them that a new society has entered the community of intelligence.'

The pair speculated as to what would be an efficient radio signal, suggesting pulses with prime numbers or 'simple arithmetical sums' for attracting attention as it would be obvious that they were artificial. They suggested calling signals to last for one minute, followed by ten minutes for a language lesson and then 50 minutes of information which would be a series of symbols.

The prize was indeed a glittering one. If a detection was made, the scientists who detected it would become the most famous scientists in the world and in history. Such a discovery would be momentous, one that would change the history of humanity. There was also the additional information that aliens could provide to consider. The pair thought it might take 50 years to receive an encyclopaedia from the stars, but what information would it contain if we were able to read and appreciate it? They suggested that we could even send messages ourselves as the radio telescope being built on Earth would be able to send messages to planets 10 light years away.

They calculated that there were about 100 suitable stars within 50 light years and seven within 15 light years that should be monitored. Three of them, Alpha Centauri, 70 Ophiuchi and 61 Cygni, would be difficult to observe because they are set against the Milky Way with too much radio interference. Other stars, such as Tau Ceti, O2 Eridani, Epsilon Eridani and Epsilon Indi, had quieter backgrounds and were prime candidates for monitoring. Morrison and Cocconi wrote in *Nature*: 'We submit, rather, that the foregoing line of argument demonstrates that the presence of interstellar signals is entirely consistent with all we now know, and that if signals are present the means of detecting them is now at hand. Few will deny the profound importance, practical and philosophical, which the detection of interstellar communications would have. We therefore feel that a discriminating search for signals deserves a considerable effort. The probability of success is difficult to estimate; but if we never search, the chance of success is zero.' The time was ripe for a search.

Otto Struve was an important astronomer at the time. A descendant of a family of famous astronomers in Russia, he fled that turbulent country in 1920 and eventually reached the United States. After arriving as an impoverished refugee, he eventually established himself and took up a position at the Yerkes Observatory, part of the University of Chicago. He later became its director and also director of the National Radio Astronomy Observatory, which gave him a key position in the debate about searching for life in space. His belief in the widespread existence of life and intelligence in the universe came from his studies of slow-rotating stars, which he believed were surrounded by planetary systems. He estimated there might be as many as 50 billion planets in our galaxy: 'An intrinsically improbable event may become highly probable if the number of events is very great … it is probable that a good many of the billions of planets in the Milky Way support intelligent forms of life. To me this conclusion is of great philosophical interest. I believe that science has reached the point where it is necessary to take into account the action of intelligent beings, in addition to the classical laws of physics.'

Struve was due to give a lecture at the Massachusetts Institute of Technology (MIT) in November 1959 and, inspired by Cocconi and Morrison's paper, changed his topic at the last minute to the search for life in space using radio telescopes. He revealed something sensational. In the coming months, the newly established National Radio Observatory's 26-metre Green Bank telescope would conduct the first search for alien signals from Sun-like stars. The targets would be Tau Ceti, 11 light years away, and Epsilon Eridani, which was only slightly closer. His lecture was reported in *The New York Times* on 4 February 1960, in a piece entitled, 'Contact with Worlds in Space Explored by Leading Scientists'. The newspaper also mentioned some questions an astronomer called Frank Drake wanted to put to aliens. He wanted to know how 'to prevent cancer and heart disease; how to prolong life; how to control the energy of the fusion process in the hydrogen bomb for industrial power; how to develop man's creative potential; and, above all, whether – and if so how – they had managed to build a culture at peace in which each individual lived a full physical and spiritual life.'

Frank Drake had an interest in using radio telescopes to look for alien life that was independent of Cocconi and Morrison's *Nature* paper. He had been convinced of the existence of aliens ever since his 1930s Chicago childhood. He later wrote: 'I could see no reason to think that humankind was the only example of civilisation, unique in the universe.' Drake had enrolled at Cornell University on a US Navy Reserve Officer Training Corps scholarship. He attended a lecture about life in space by Otto Struve in 1951. After college, he served as an electronics officer on the heavy cruiser USS *Albany* and then went to Harvard to study radio astronomy. He had already cut his teeth using the 18-metre radio telescope at Harvard. He had pointed it at the well-known Pleiades star cluster in the constellation of Taurus and got a strong signal. For an instant, he later wrote, it got his heart racing only to be disappointed when the signal persisted as the radio telescope was pointed elsewhere in the sky.

On 3 April 1960, *The New York Times* reported that Project Ozma was ready to go. The project was named after the queen of L. Frank Baum's imaginary land of Oz – a place 'very far away, difficult to reach, and populated by strange and exotic beings'. Five days later, in the early hours of a cold and misty morning,

29-year-old Drake and two student assistants, Margaret Hurley and Ellen Gunderman, took control of the Howard E. Tatel Radio Telescope at Green Bank and steered it towards Tau Ceti, tuning the receiver to the 21-centimetre line. Drake was embarking on what would be a major role in the search for intelligent life in space. 'What we were doing was unprecedented, of course, and no one knew what to expect. Even I, in my fever of enthusiasm, couldn't assume that we would really detect an intelligent signal,' he later reflected.

Nothing was detected from Tau Ceti. At noon, he moved to Epsilon Eridani and within a few minutes recorded a strong, pulsed signal. Drake stared at the pen and ink chart recording, which was accompanied by a series of dramatic booms from the loudspeaker – exactly what he expected to hear from an extraterrestrial civilisation trying to attract attention.

PALIMPSEST

It all sounds very hopeful. Humanity's connection with the cosmos accepted with love and reverence, as Tesla had put it. But there is a darker side. What might have been coming out of the loudspeakers in the receiving room at Green Bank? Did we really have any idea what we might be doing?

We do not know what life there is out there, if any at all. If we received messages, what would they be like? They could come from all directions and not just in radio waves. They could be traversing space, passing through each other unhindered, disturbances of photons, ripples in space time, perhaps patterns written in quantum-entangled particles, superimposed on one another, fading with the inverse square. All hoping to be detected. Faint traces of individual voices, cries and calls, like distant sounds in a night-time jungle, together forming a sentient background radiation and a cosmic palimpsest. Each message a civilisation's mark on the cosmos, a reflection of strange lives and experiences beyond any single entity's comprehension. Artificial intelligence programmes moving between star clusters, finding fertile havens in many environments. Holographic art forms passing between species. Souls moving at lightspeed. The obvious and the unfathomable.

Maybe some of these messages are from a young species like ourselves who have just learned how to send messages to the stars? Some may be from those who have been around since the universe was young. Will they be wise, showing the way for others to survive, unlocking the door to secrets it took them aeons to discover? Some might say they survived, or perhaps endured, against setbacks, telling others that they could do the same. All of them would be saying 'we are here'.

But think on this: could these messages be from the rampant and the shattering, those that hurricane-like ravage planets? Would we want to know them? Some may excel in evasion and counterpoint, some deliver tension. Could it be that in the dark is the sinister, the life that will tear to shreds any competition for whom regret and hope are themselves alien? And all of this booming on the Project Ozma loudspeakers.

I wonder about such messages especially when I look at the stars. Would we ever be able to read the spellbound poetry of the outer reaches of our galaxy, witness the oppression of its inner bulge or the simple strength of its spiral arms? There could be the Dyson spherics, the black hole colonists, those who live in the slipstream of relativity and in the folds of space time. The last cry of those whose societies faltered or those denied oblivion. Those being destroyed by radiation. Psalms from dying and long-dead authors. Those who survived, those who connected, those who merged, those who absorbed, those who destroyed with terrible splendour. The messages may come from the perturbed and the frightened, from the half-formed and from those that span the stars. A brief rail against life's exquisite dissonance against entropy. All ask: what is permanent about the universe? What is there to be relished? What else is possible? Who is there? Know us.

There were no alien signals sounding from Ozma's sound system. It had detected a strong signal, too strong, and although it was never tracked down satisfactorily there are some who think it came from a U-2 spy plane. Since then, there have been many searches using radio telescopes. Long hours moving from star to star, and there have been more false alarms.

The early 1960s was a hopeful time in the search for aliens. Many scientists thought there was a strong possibility that the sky was awash with signals from them and that they would detect aliens, if not immediately, then very soon. The limited nature of Project Ozma and its failure proved little, and Frank Drake was undaunted. 'At this very minute, with almost absolute certainty, radio waves sent forth by other intelligent civilisations are falling on the Earth. A telescope can be built that, pointed in the right place, and tuned to the right frequency, could discover these waves,' he said. Cocconi and Morrison had shown that even we, crude and unsophisticated as we were, had the means to signal across interstellar distances. How much better could aliens do if they had been around much longer than humans? Was it just a technical problem involving the right size of radio telescope, the right frequency and pointing it in the right direction at the right time? But, sixty years after these first attempts, no such signal has been detected. Why?

In 1961, the day after Halloween, 31-year-old Drake joined 26-year-old Carl Sagan, Philip Morrison, Otto Struve, businessman Dana Atchley, dolphin researcher John Lilly, biochemist Melvin Calvin, astronomer Su-Shu Huang and vice-president of research at Hewlett-Packard Bernard Oliver, along with J. Peter Pearman at a conference at the Green Bank Observatory. Pearman, an official at the National Academy of Sciences, had been fascinated by Project Ozma and organised this follow-up meeting. Calvin was just about to receive news that he had won a Nobel Prize for his work on photosynthesis. John Lilly called the participants the 'Order of the Dolphin'. Sagan devised an acronym for the topic, CETI – Communication with Extraterrestrial Intelligence. CETI soon gave way to SETI with 'Communication with' replaced by 'Search for'.

Despite the hope that alien contact could be imminent, there was a feeling that SETI needed a firmer footing in astronomy and not just on the realisation that radio telescopes could send and receive signals over interstellar distances. They were concerned by comments made by the Italian–American physicist Enrico Fermi, who at lunch one day several years earlier, when the topic turned to aliens, asked, 'Where is everybody?' By now, the Milky Way should have been colonised by aliens as even if they travelled from star to star rather slowly they should have had enough time to traverse the whole galaxy. They knew they had no satisfactory response and that speculating about alien life was one thing but actually searching was another, and to search they needed to persuade less enthusiastic astronomers, those who granted telescope time. Was there any way to put some numbers into the search, to calculate what might be the chances of detecting aliens?

Frank Drake proposed what became a famous equation to estimate the number of communicating civilisations in our galaxy. It was a way to put some useful numbers into the argument, adding scientific credibility as well as a way of justifying and optimising searches. By breaking down what was a great unknown – the number of communicating aliens out there – into a series of smaller questions, it made the search seem more realistic, more promising. Drake's Equation produced the value N as the number of communicating civilisations in the galaxy at any particular time. The first section of the equation deals with the rate at which stars are born, and in particular those that will live long enough to have a planet with life. The factors to be considered include the fraction of stars with planets, the fraction of habitable planets where life develops and the fraction of those with intelligent life. But this was not a process of rigorous science, more like informed guesswork, which in science is sometimes all that is possible. But it became even more speculative as later factors were reached. The final two factors to consider are the fraction of civilisations that communicate over interstellar distances and the length of time they do so, basically sociological considerations of which we are entirely ignorant. At the end of this calculation, what was left was the number of communicating aliens out there in the galaxy.

Clearly, the Drake equation is limited and has no real predictive capability, but those at the Green Bank conference, sitting in the shadow of the Project Ozma telescope, were optimists. Even though they recognised the enormous uncertainties involved, they concluded that there were between 1,000 and 100 million alien civilisations communicating in the galaxy at that point in time. This is a very wide range. If it was 1,000 we as humans would have little chance of detecting them, if it's 100 million then perhaps we would have a chance. If there were 100 million aliens, they could be found at about one in 1,000 stars, with an average distance of 30 light years. The two stars searched for by Project Ozma were about 11 light years away. Over such a cosmically short distance we could exchange signals in a human lifetime and conceivably travel there in a few millennia with interstellar spacecraft. The optimism among those at the conference rose. At the end of the meeting, Otto Struve broke open a bottle of champagne and toasted: 'To the value of L [the lifetime of an alien civilisation]. May it prove to be a very large number.'

Unsurprisingly, the Drake equation was not universally liked; some said it was just a way of organising our ignorance. According to Bernard Lovell, it was 'the most unequation-like equation I had ever come across', and, he added, 'if any of my students had submitted it to me I would have failed them'. The equation is biased, certainly, but then so is much in the SETI endeavour. Over the years, different scientists have inserted different numbers for each factor in the equation and reached wildly different conclusions, following a changing fashion for aliens. Three decades after it was first formulated, Sagan reduced the incidence of aliens a hundredfold. There are about as many different opinions about the Drake equation as there are number of people who have studied it. At the time when Sagan was going through a pessimistic (for him) phase, science fiction author Isaac Asimov was saying that there might be 400 million planets in our galaxy and that nearly all of them would be home to an alien civilisation that was far more technologically advanced than ourselves. With such an equation you can support many views. Scientifically, the logical thing to do was to realise how uncertain the factors were and seek to experimentally verify N, that is perform the search for alien life. As Morrison said in *Nature*: 'If we never search, the chance of success is zero.'

Beneath the surface of the Green Bank meeting was the worry that an alien message would be impossible to understand. After all, the Linear B script of the Ancient Minoans in Greece had withstood generations of attempts to decipher it, and it was a human language! Egyptian hieroglyphics required the Rosetta Stone to be understood. Calvin even suggested that we might have to find two alien civilisations talking to each other to understand what was going on!

To illustrate the message problem, a few months after the meeting, Frank Drake sent the participants a letter containing a typed sheet of paper with a long string of 0s and 1s, which Drake said was an example of a message that might be received from another civilisation in space. There were 551 digits, but what did it mean? Only Bernard Oliver was able to solve it. It was a representation of a picture. The figure 551 is a prime number and is the product of 19 × 29. Therefore, formatting the message into 29 rows and nineteen columns reveals an image of a pixilated four-legged creature and a diagram of a planetary system. Oliver was impressed and replied in the same fashion. When Drake decoded the reply, he saw a picture of a martini glass.

Despite the desires of the Order of the Dolphin, there were few searches conducted in the 1960s when it was the Soviet Union that dominated SETI. Rather than searching the vicinities of nearby stars, the Soviets were using omni-directional antennas to observe large chunks of the sky in the hope that there would be a few very advanced civilisations out there capable of radiating enormous amounts of transmitter power. According to American scientists, this was a bad strategy.

But then on 12 April 1965, TASS – the official Soviet news agency – sent out a story entitled, 'We are signalled by a friendly civilisation'. Frank Drake sent a telegram to the scientists concerned, Nikolai Kardashev and Iosif Shklovsky, congratulating them and asking for more information. The 'signal' was a radio source designated CTA-102 that changed in radio intensity and was interpreted as coming from an intelligence in deep space. Radio sources that changed in strength were hardly known at the time, although Bernard Lovell had been pioneering a series of observations in the late 1950s at Jodrell Bank, looking at explosions on the surface of nearby so-called flare stars. The Soviets were thrilled and replied: 'We have a matter which may be one of the most outstanding discoveries in the history of radio astronomy.' It came at the same time as Yuri Gagarin became the first man to go into space and added to the Soviet Union's prestige and superiority over the United States. The problem was that it wasn't true.

Something changed and quickly. Kardashev was quoted as saying 'a supercivilisation is discovered', and Shklovsky declared himself 'excited'. But at a subsequent press conference, Shklovsky was dismissive of the claim and attacked TASS for misquoting him. He even received a telegram from the California Institute of Technology informing him that it wasn't aliens. In the British press, Lovell was quoted as saying the Soviet discovery was sad.

Shklovsky clearly wanted the signal to be aliens, he told me so many years later. At 49 years old, he had made a distinguished contribution to Soviet science, even though he resented the extent of state control over the discipline. He made discoveries about the Sun's hot outer atmosphere and speculated about the origin of the mysterious cosmic rays detected coming from deep space. He was highly regarded for his suggestion that the radiation from the Crab Nebula was due to so-called synchrotron radiation, in which unusually energetic electrons spiral through magnetic fields at speeds close to that of light. He speculated that cosmic rays from supernova explosions within 300 light years of the Sun could have caused mass extinctions of life on Earth.

He also had a soft spot for alien life. A few years earlier, he said that the orbit of Mars' inner satellite Phobos was decaying, and from it calculated that Phobos must have an exceptionally low density, and it might in fact be hollow and artificial. We now know, as we have always really known, Phobos is a chunk of rock. He later sent a book about searching for life in space to Carl Sagan, who added bits of his own text producing what is regarded as a classic work on the topic of searching for intelligent life in space.

Shklovsky allowed himself to be carried away for a little while, after all he, and the Americans, really expected to find intelligent life in space and rather easily at that. CTA-102 proved to be a distant so-called quasar, the components of which did vary in intensity. Its discovery led Kardashev to speculate on what would happen if a supercivilisation had access to enormous power. We will return to this later.

In the 1960s, the USSR had a small yet diverse group of scientists interested in SETI, listening for signals from individual stars as well as performing whole sky sweeps. Their RATAN-600 radio telescope was constructed at the time, partly with SETI in mind.

In the late 1960s, John Billingham, who worked at NASA's Ames Research Center, tried to get NASA interested in SETI. Billingham was a trained medical doctor who had worked on the liquid-cooled inner garments for the Apollo spacesuits. Consequently, NASA started to think about the technology required for what they considered an effective search. They brought in a team of outside experts alongside Billingham under the direction of Bernard Oliver, who took leave from Hewlett-Packard. They produced a study called Project Cyclops. When I was a young radio astronomer at Jodrell Bank, I remember going through Project Cyclops as from time to time we were visited by those who were involved in it. It was without doubt the most unfeasible suggestion that SETI had ever seen, with its great number of purpose-built large radio telescopes and its $10 billion price tag. Technically, it did not fare well as within a relatively short time advances in receiver technology obviated the need for such large radio dishes with the stringent specifications required. The project showed that SETI was going down a one-way street; the ideas were big but the big ambitions behind them were never going to happen. Cyclops was an all or nothing gamble. SETI got nothing.

In the mid-1970s, an astronomer called Michael Hart gave the astronomical community pause for thought. In an impressive and influential paper published in the *Quarterly Journal of the Royal Astronomical Society* entitled, 'An Explanation for the Absence of Extraterrestrials on Earth', he concluded that SETI was a waste of time and money. He pointed out that there are no aliens on Earth now, a statement that he referred to as 'Fact A'. 'Fact A, like all facts, requires an explanation,' wrote Hart. He went on to conclude that Fact A implies that intelligent life from outer space does not exist. In other words, there is no evidence that aliens have been on Earth, therefore he decided we are alone in the galaxy.

In 1975, NASA funded some studies through Philip Morrison at MIT and in the same year Ozma 2 was conducted, using the largest radio telescope at the Green Bank Observatory, the site of the original search. Such had been the development of electronics in the meantime that in just one minute the new search collected as much data as the original Project Ozma could have done in nineteen years. A series of Sun-like stars were examined, nothing unusual was seen. Clearly, the initial optimism that aliens would be found swiftly was evaporating, which led to speculations that at the least we might be living in a quiet neighbourhood. The alternative, that it really was a waste of time and money, was countered by the 'if we don't look we will never know' argument, but something had changed and were it not for the championing of the subject by Carl Sagan, for whom it was a dominant theme of his life, the search for alien life might have been pushed to the margins more than it was.

A negative search conducted in 1976 by the University of California at Los Angeles by astronomers Ben Zuckerman and Pat Palmer left both pessimistic. 'I guess I was an agnostic in the matter of intelligent life in the universe when we first started those searches. But in the late '70s my view started to shift. I felt more and more that they weren't out there and that there might be very few or no civilisations in our Milky Way more advanced than ours,' said Zuckerman, who was swayed as much by the views expressed by Michael Hart as by his negative results.

The longest continuous series of SETI observations were carried out by the radio telescope at Ohio State University between 1973 and 1995. On 15 August 1977, their radio telescope, nicknamed 'Big Ear', detected what many believe is the best candidate for an alien signal recorded to date. It came from the direction of the constellation Sagittarius, roughly the centre of our galaxy. Big Ear is not a steerable dish but a fixed curved reflector of steel mesh and a tilting flat reflector. It can scan in an up–down direction but allows the Earth's rotation to move celestial radio sources through its fixed field of view, which means that when a signal passes through it it produces a symmetrical curve.

Astronomer Jerry Ehman saw the signal a few days after it had been detected when reviewing printouts. He noticed one particular series of numbers and letters – 6EQUJ5 – that in the computer code represented the most powerful signal he had ever seen. He wrote next to it 'Wow!'. It was a strong and transient signal. Big Ear's design had a dual-horn feed meaning that radio sources would first appear in one horn and then a few minutes later in the other but, as Ehman looked at the printout, he could find no sign of the signal in the other horn. The signal had ceased before the second horn would have been able to pick it up.

Later, the much more sensitive Very Large Array system of radio telescopes in New Mexico looked at the coordinates but detected nothing. Ehman said: 'We should have seen it again when we looked for it 50 times.' But nothing has ever been detected from that direction since. Has the signal, if it was that, been repeated at times when we weren't looking? Or will it be repeated on timescales longer than human lifespan? It is all speculation and Ehman did not want to draw 'vast conclusions from half-vast data', suggesting the signal may have been military in origin. That might be the case but the 'Wow!' signal remains unexplained and almost certainly always will be. Nobody using a radio telescope looks at those coordinates any more. By the late 1970s, NASA's SETI effort became established at the Jet Propulsion Laboratory (JPL) in Pasadena, California. The NASA Ames Research Center looked at 1,000 Sun-like stars and JPL conducted sweeps in all directions in a sky survey.

Trouble was brewing in Washington DC. By 1978, although SETI was receiving a low level of funding it did come to the attention of William Proxmire, a Democratic senator from Wisconsin in post since 1957, when he took the seat from the disgraced Joseph McCarthy, who had just died. He was re-elected five times and hardly ever spent any campaign money. He became famous for his 'Golden Fleece Awards' for any government department that in his view wasted money. The Environment Protection Agency got one for spending over $1 million on preserving a New York sewer as a historical monument. NASA was awarded the fleece for wasting money on SETI. It was a sign of things to come.

In 1980, Frank Tipler, a 33-year-old physicist then at the University of Texas expanded on Michael Hart's views again, writing in the *Quarterly Journal of the Royal Astronomical Society* entitled, 'Extraterrestrial Intelligent Beings do not Exist'. Since aliens aren't on Earth, we are likely alone, Tipler reiterated. 'The point is that a belief in the existence of extraterrestrial intelligent beings anywhere in the galaxy is not significantly different from the widespread belief that UFOs are extraterrestrial spaceships. In fact I strongly suspect the psychological motivations of both beliefs to be the same, namely "The expectation that we are going to be saved from ourselves by some miraculous interstellar intervention."' Tipler's views were unpopular among the SETI community perhaps because they unconsciously touched a nerve: there was no difference between a scientist who wanted to ask aliens questions about philosophy and a cure for cancer, and those who looked for guidance from the purported alien spaceship flying around in their backyard.

In 1982, a statement signed by 70 prominent scientists from fourteen countries urged a 'coordinated, worldwide and systematic search' be conducted. It was organised by Carl Sagan, who was flush with fame as a result of his internationally successful TV series *Cosmos*. It said:

> The human species is now able to communicate with other civilizations in space, if such exist. Using current radioastronomical technology, it is possible for us to receive signals from civilizations no more advanced than we are over a distance of at least many thousands of light years. The cost of a systematic international research effort, using existing radio telescopes, is as low as a few million dollars per year for one or two decades. The program would be more than a million times more thorough than all previous searches, by all nations, put together. The results – whether positive or negative – would have profound implications for our view of our universe and ourselves.

The call for funding had to address the obvious criticisms: haven't we all heard this before? Why haven't they been found already? In anticipation of such opposition, the statement argued: 'It has been suggested that the apparent absence of a major reworking of the Galaxy by very advanced beings, or the apparent absence of extraterrestrial colonists in the solar system, demonstrates that there are no extraterrestrial intelligent beings anywhere. At the very least, this argument depends on a major extrapolation from the circumstances on Earth, here and now. The radio search, on the other hand, assumes nothing about other civilizations that has not transpired in ours.' It ended: 'We urge the organization of a coordinated, worldwide and systematic search for extraterrestrial intelligence.'

In November 1982, the US Congress approved a modest $1.5 million in NASA's budget for the search for extraterrestrial life. Ben Zuckerman later said: 'I signed the petition in the journal *Science* that Sagan drew up because although the negative arguments are strong, I don't think they are airtight, we still have to do more searching. The searches up to now have not been sensitive or comprehensive enough that we can give them a great deal of weight. It would be nice to get that extra radio data.' At the same time, Greek astronomer Michael Papagiannis said: 'I think that in the next ten or twenty years we will either have detected some sort of signal from SETI, or we will begin to accept the fact that we are alone in our galaxy.'

To get the extra radio data, Zuckerberg wanted a truck loaded with wooden crates to make the 700-kilometre journey south and east from NASA's Ames Research Center to the 26-metre dish of the JPL's Goldstone Observatory in California's

Mojave Desert, transporting a VAX mainframe computer, circuit boards, copper cables and a vacuum bottle containing liquid helium to cool the radio receiver. The aim was to split the output from the telescope into 74,000 narrow frequency bands and search for something unusual. The hope was that it was the first of a long-term, reasonably well-funded campaign. John Billingham, head of NASA's Extraterrestrial Research Division at Ames, said: 'The way it is structured now, we have a five-year R&D program in which to check out exhaustively the design of the SETI systems … Then we will have roughly a ten-year period during which we will build the search machine proper.' A year later, the programme was cancelled in a move instigated by Senator Proxmire.

Unsurprisingly, the author of the *Science* letter and public spokesperson for all things SETI, Carl Sagan, went to see the senator. Proxmire had voted against the Vietnam War and hated the prevailing concept of mutually assured destruction by nuclear weapons. These were views that Sagan shared. If we find a signal from an extraterrestrial civilisation, Sagan argued, it would suggest that intelligent beings do not inevitably destroy themselves and there is hope for humanity. Proxmire changed his mind, actually apologised and the following year funding for SETI was reinstated.

Starting in 1983, radio astronomers conducted searches, including the Planetary Society's Project META (Megachannel Extraterrestrial Assay), which was a continuous search partly funded by movie director Steven Spielberg. Its results were published in 1993 and it detected 37 candidate signals, all located in the galactic plane that could not be explained, but they were never seen again so we cannot know what they were. Even though the level of funding was relatively low, it was adequate. Gaining confidence, the private, non-profit, SETI Institute was established in California to provide a focus, as many believed NASA diluted the search by looking for primitive life in the solar system. The Planetary Society also provided some funds for two researchers to go to NASA's Deep Space Network radio telescope in Australia to carry out a limited series of observations. A 'Suitcase SETI' portable system was developed that eventually became Project Sentinel and then the Megachannel Extraterrestrial Assay. At the same time, Shklovsky and Kardashev were performing searches from the USSR.

In 1988, NASA formally endorsed SETI again and found more money for the programme. SETI scientists were eager to begin the search. They were anxious not only because of their lack of success so far but also because the radio spectrum was becoming cloudy as a result of interference and the demand for frequencies for new commercial applications.

In 1990, the Bush administration requested $12 million for SETI to start what was to be called the Microwave Observing Project (MOP), but there was opposition. Republican Congressman Ronald Machtley said: 'We cannot spend money on curiosity today when we have a deficit.' He went on to add: 'We have no, and I repeat no scientific evidence that there is anything beyond our galaxy. If, in fact, there is a superintelligent form of life out there, might it be easier just to listen and let them call us?' This was exactly what NASA was intending to do! Republican Silvio Conte's criticism was more barbed and asked why one should spend millions of dollars looking for aliens when one could spend '75 cents to buy a tabloid at the local supermarket', which is full of aliens. These attacks notwithstanding, SETI came away with $11.5 million in funding. It had survived, for now. NASA decided to respond to the ridicule by reorganising the project, MOP became the High Resolution Microwave Survey (HRMS), which sounded a bit more scientific. Still, the House Committee on Appropriations tried to cancel the project and it was saved by Senator Jake Garn, who spoke about the new perspective he gained from his recent flight on the Space Shuttle and his belief that searching for life in space was a religious quest.

So, in 1992, on the 500th anniversary of Columbus' arrival in the New World, the observations began after a ceremony daubed with optimism and relief. At an event at Arecibo, Puerto Rico, John Billingham stepped up to give an opening speech. 'This is the beginning of the next age of discovery,' he said. 'We sail into the future, just as Columbus did on this day 500 years ago. We accept the challenge of searching for a new world.' Simultaneously with Arecibo, NASA's Goldstone telescope in California, started the survey portion of the search. Arecibo was aimed at the star GL615.1A, 63 light years away in the constellation Hercules and one of 24 Sun-like stars selected for the initial observations. *The New York Times* reported: 'There was speculation as to what future archaeologists might surmise if they happened on the ruins of these stone pillars, aluminum panels and huge steel cables and girders. Here a society with scientist-priests communicated with their gods in the heavens? Some Columbuses sought the cosmic Indies, never found? Or this was the place where humans listened in the jungle stillness and for the first time heard that they are not alone in the universe?'

It was to be a ten-year, $100 million observation programme. HRMS was to have two components: a targeted search of nearby stars using the Arecibo telescope; and an all-sky scan using the dishes of NASA's Deep Space Network. It went well for a year, and then it happened again. During NASA budget hearings, Republican Senator Richard Bryan addressed Daniel Goldin, the politically inexperienced relatively new administrator of NASA: 'Mr. Goldin, something in your budget doesn't pass the smell test.' He was referring to SETI. In September 1993, Bryan tabled a last-minute amendment to kill the programme and the Senate concurred. He issued a press statement: 'As of today, millions have been spent and we have yet to bag a single little green fellow. Not a single Martian has said "Take me to your leader", and not a single flying saucer has applied for FAA [Federal Aviation Administration] approval. It may be funny to some, except the punchline includes a $12.3 million price tag to the taxpayer.' Bryan then said: 'This hopefully will be the end of Martian hunting season at the taxpayer's expense.' As far as NASA was concerned, SETI was dead.

In response, Bernard Oliver wrote: 'Millions of transistors, memory cells, and other high-tech products of our ingenuity have been woven into a brain whose sole aim in life is to detect and verify the origin of tiny signals – less energetic than the smallest atomic particle – that have crossed the light years we cannot. Such signals will tell us that we are not alone, that the astonishing process that has produced us out of the fiery furnace of the Big Bang has also occurred elsewhere. Lo, from that single fact, all our philosophy would be enriched ... To save the American Taxpayer about eight cents per year, we are to be denied the chance to explore the universe and the sentient life forms that fill it.'

Why did it happen? Why did, in what was in NASA's grand scheme a very small project, taking up 0.1 per cent of its budget, get cancelled when it could have, like so many other projects at a similar level, survived under the radar? Perhaps because of its high-profile subject, SETI seemed bigger than it was. At the same time, the big physics Superconducting Super Collider project was cancelled. This project was clearly controversial, overblown and badly managed, so perhaps it rubbed off on SETI, unfairly. In addition, NASA was having a very bad time barely winning budget battles over the planned Space Station programme and solid-fuelled rockets. The problems with the misshapen mirror of the Hubble Space Telescope only added to the agency's woes, as did the loss of NASA's multi-million-dollar Mars Observer spacecraft that failed before it entered the orbit of Mars. With such forces at play, SETI and its giggle factor was bad politics. Suddenly the only people who supported it were those working on it. I understand Carl Sagan tried to meet Senator Bryan, as he had met Senator Proxmire a decade earlier, but Bryan refused to see him. Bryan's office once called NASA only to find there was no one at NASA able to take his call. Afterwards, John Gibbons, President Clinton's science adviser, said: 'We've done a lot of listening [for alien signals] already, and if there was anything obviously out there, I think we would have gotten some signal.'

Looking back, SETI was an easy project to cancel and few ran to its defence. Despite having superstar Sagan and his influence, which in the early 1990s was probably at its peak, it all fell apart rather swiftly. Not all scientists supported SETI and contractors saw little prospect of making money from it. One SETI scientist analysed the constitution of the various committees involved, finding that one contained more former undertakers than former scientists. UFOs and bad timing did for it as the Republicans fought President Clinton's executive to balance the budget by looking for anything they could cut. When the government was shut down because a budget could not be agreed, no one cared about aliens.

SETI personnel reacted quickly. With sponsors that included William R. Hewlett and David Packard, Intel's Gordon Moore and Microsoft co-founder Paul Allen, the SETI Institute secured $7.5 million to cover costs for the next few years. From the ashes of expectations, planning for Project Phoenix began, which concentrated on the targeted radio observations and abandoned the all-sky survey. Eventually, Project Phoenix listened carefully to 800 Sun-like stars mostly closer than 240 light years. Nothing was found. In 1996, the SERENDIP (Search for Extraterrestrial Radio Emissions from Nearby Developed Intelligent Populations) initiative used the Arecibo telescope every 1.7 seconds on 168 million radio channels to look for a signal. Reflecting on the project, scientist Dan Werthimer said: 'I'm optimistic in the long run. The Universe is probably teeming with life. If we're lucky we'll find something in our lifetimes.'

Searches are ongoing, but until recently were at a very low level with barely a handful of astronomers carrying out sporadic observations comprising a dozen or so hours of telescope time. For a while, the SETI Institute collaborated with the University of Berkeley to develop an array of small radio telescopes based on commercially available satellite dishes initially called the One Hectare Telescope and later renamed the Allen Telescope Array. But in 2008 the financial crisis struck and shortly afterwards Berkeley withdrew from the project, which led to it being put into hibernation for eight months. The planned expansion from 42 to 350 dishes never happened and the unmown lawns at the site testify to its lack of money. It carried out regular searches between 2007 and 2015, but nothing conclusive was found. SERENDIP continues to this day. Nothing yet.

In 2010, fifty years after Project Ozma, astronomers got together to organise Project Dorothy, a two-year scheme, using different kinds of telescopes, including Green Bank, in fifteen countries to look at seven nearby stars, including Tau Ceti and Epsilon Eridani. Nothing was seen. Frank Drake commented: 'It is thrilling for me to witness the beginnings of Project Dorothy, the continuation of my search of fifty years ago. To have so many talented people using so many telescopes in this new search, with the electronics and computer equipment of today, is a joyful thing to me. The equipment of today is far better than what we could have fifty years ago, and will result in both very much better and very much more data than could be obtained then … Over the past fifty years our searches have not yet produced the discovery we all hope for. This is understandable – in our vast and awesome universe it will take long, painstaking, and comprehensive searches before we will have a good chance of success. This is the major lesson learned from previous searches.' Long, painstaking indeed. The heady optimism of the Order of the Dolphin was long gone. Where were the aliens?

What was needed was new funds and in a technological landscape populated by the likes of Elon Musk and his SpaceX company and Jeff Bezos with Blue Origin and their shared dreams of space exploration, those funds arrived in the shape of another billionaire. The new funds have also been 'a huge catalyst' for training scientists in SETI, says Jason Wright, director of the Penn State Extraterrestrial Intelligence Center, which opened in 2020 and is dedicated to looking for alien technosignatures – signs of alien technology or influence.

Named after Yuri Gagarin, Yuri Milner was born and educated in Moscow and worked as a particle physicist at the Lebedev Physical Institute. Amid the collapsing USSR, in 1990 he left to study business at the University of Pennsylvania and founded an internet investment fund. He says he made some lucky early investments, such as in Facebook, Twitter, Spotify and Airbnb. He became aware of the dire condition SETI was in and offered $100 million to fund his 'Breakthrough Listen' initiative, an amount which seems like small change for someone worth $3.8 billion. I remember going to the launch press conference in London where Milner was flanked by Sagan's widow Ann Druyan, Stephen Hawking, planet hunter Geoff Marcy and Frank Drake.

This is clearly the most comprehensive search yet carried out. It began in January 2016 and is set to search for ten years. Controversially to some, it is not based at California's SETI Institute but at the west-coast University of Berkeley. One third of the money in the project is being used to purchase telescope time at the familiar Green Bank Observatory and the Parkes Observatory in Australia where it has signed contracts for 20 and 25 per cent of their available time respectively. Breakthrough Listen is looking at 1 million nearby stars and the centres of 100 galaxies. It is reported that the system could detect Earth-leakage-like signals from about 15 light years and can detect a transmitter equivalent to an aircraft radar from the nearest 1,000 stars. In October 2016, the newly built Five-hundred-meter Aperture Spherical Radio Telescope (FAST) in China joined in. This telescope is of the Arecibo-type design that is built into the landscape and is currently the world's largest radio telescope.

In December 2020, it was reported that in April and May of 2019, a narrowband signal at 982.002 MHz was detected that displayed variations in frequency consistent with the movement of a planet around a star, in this case the closest star to the Earth – Proxima Centauri. No changes in the strength of the signal were detected and it's been designated 'Breakthrough Listen Candidate 1'. At the time of writing, the researchers are yet to exclude terrestrial interference, but they don't think it's aliens.

Breakthrough Listen is expanding its observing base by adding MeerKAT, a South African array of 64 dishes, each 13 metres across. Instead of buying time, it is sampling its data stream while the telescope carries out its normal observations, just like SERENDIP. The Karl G. Jansky Very Large Array in New Mexico, the workhorse of US radio astronomy, will also be joining in. All of the data from Breakthrough Listen will be available to the public – 10 gigabytes per second – the largest amount of scientific data ever made available to them. 'We will be able to generate as much data in one day that would have taken previous SETI searches one year,' Milner said. 'The hardware and the software used in the Breakthrough project will be compatible with other telescopes around the world, so they too can search for intelligent life,' he added.

As an aside, piggy-backing radio observations is far from new. As a young radio astronomer at Jodrell Bank in the late 1970s and 1980s, I obtained my PhD by using the main telescope along with one 200 kilometres away to look for explosions on the surfaces of nearby stars, a continuation of the observing effort Bernard Lovell mentioned in his reply to Cocconi and Morrison. We were looking for flares, so we had the telescope set to parcel the data into 30-second chunks, although we were not observing with a very narrow frequency range. We saw flares from some of the nearest red dwarf stars in the solar neighbourhood. It occurred to me that I should look at the data for anything that might be suspicious alien-wise. So, I wrote a few small computer routines to analyse the data but nothing was seen. On another occasion, I took part in an observing run looking for what is called Hawking radiation, which we will discuss later. Briefly, small black holes made when the universe was young, if they exist, should about now, 14 billion years later, be disappearing in a flash of radiation, including radio waves. So, we used the Lovell dish to look towards the centre of our galaxy and in the opposite direction, as well as up and down from the galactic plane and up and down our local galactic spiral arm. We found nothing at our level of sensitivity, either black hole flashes or alien signals. Otherwise, this would have been a very different book!

There is another way to send signals between the stars. Because of bias and fashion, this method hasn't received its fair share of attention from those who prefer radio waves in their search for signals from space. This is the prospect that aliens might be sending laser signals towards us. When the laser was invented in 1960, during the time when radio searches were becoming the big thing, it was dismissed as a curiosity that was barely able to produce a few milliwatts and something that would find few practical applications. Today of course, lasers can be found everywhere. Using high-powered lasers, it is possible to swiftly transmit large amounts of data across interstellar distances.

Today, the most powerful laser in the world has a power output of about 10 petawatts in pulses lasting just a trillionth of a second. Chinese researchers are aiming to create a 100-petawatt laser in the near future. Its single pulse would be 10,000 times more powerful than all the electricity grids in the world combined, though of course lasting for a very brief time. Because they would stand out as being very different from starlight, our lasers are already bright enough to be seen across interstellar distances, and we even have detectors that can easily register pulses of just a billionth of a second. In the right circumstances, it would be possible to transmit the entire content of the US Library of the Congress in just a few minutes using lasers.

Lasers have other advantages as a medium for sending messages. They don't suffer from a smearing of frequencies like radio waves, and although their beams broaden when beamed over interstellar distances, this spread – perhaps several hundred million kilometres – would be useful as it would bathe more than one possible planet with its light. As one would expect, there are some problems with using lasers, particularly when it comes to interstellar dust, but lasers tuned to infrared light can overcome such issues. In fact, the inventor of the laser, Charles Townes, realised that the universe was transparent to infrared lasers up to about a few hundred light years, but that could be enough.

Few were interested in lasers for interstellar communication except Stuart Kingsley, who from the 1970s, with disparagement from the radio optimists, carried out observations using his own telescope. On one occasion at a meeting in 1990, he was invited to speak at NASA's Ames Research Center and was told just before his talk not to say anything about the superior beaming capabilities of a laser. He ignored the advice and his argument was persuasive. With radio drawing a blank, even Frank Drake became interested in lasers leading a search in 1999 from the Lick Observatory in California examining 5,000 stars for flashes of laser light over a decade. Nothing was found. Today, lasers, especially infrared ones, have just as much justification behind them as tools of alien communication as radio waves and perhaps in the future the way to go is to look for short radio pulses and especially infrared laser bursts. There is a project called NIROSETI (Near Infrared Optical SETI), which is promising, if only it could be funded properly. Perhaps there is a message waiting to be detected sent by aliens from a nearby planetary system who have tailored its properties to pass through interstellar dust and even through the possible atmosphere of a water-rich world. With the technology we have, we can outshine our Sun by thousands of times over thousands of light years. We may soon be doing this.

Some believe that the Breakthrough Listen initiative is reducing the giggle factor and bringing SETI into the mainstream for reasons other than money talks. Almost three decades after Senator Bryan, in 2018, NASA held a workshop on the prospect of detecting alien technosignatures. It even awarded a grant along those lines – its first ever SETI-related grant not involving radio searches.

Yuri Milner has certainly been busy. In 2016, he, Stephen Hawking and Facebook founder Mark Zuckerberg established the Breakthrough Starshot project to develop a proof-of-concept fleet of tiny interstellar probes called StarChips – minute devices attached to small solar sails and propelled by a laser fired at them. If the project works, they will be aimed at the Alpha Centauri star system 4.4 light years away. After the initiative was announced, a planet was found orbiting a star in the system, Proxima Centauri, around which it orbits in the so-called habitable zone where conditions could be suitable for life. For the project to succeed, fundamental advances need to be made in over twenty fields of engineering – exceeding the current limits of technology by a long way.

The goal is to reach a speed about 20 per cent of the speed of light, resulting in a twenty- to 30-year journey to the star and four years for a return signal. The laser required to propel the StarChips would have to be a gigantic multi-kilometre phased array of steerable laser beams with a combined coherent power output of up to 100 gigawatts. Such beams will make the Earth visible at interstellar distances if any aliens happen to be looking at us at the right time.

Much has been said about humanity reaching the threshold

of interstellar communications when we developed radio tel-escopes, our so-called entry card to a potential galactic club. But what if that wasn't enough? What if there are other discoveries we make or technologies we develop that are better thresholds. Could it be that developing radio telescopes isn't in itself enough and that aliens are waiting for us to pass further thresholds to open a new window to contact? We may just have done that.

RIPPLES

'I know perfectly well that at this very moment the whole universe is listening to us – and that every word we say echoes to the remotest star.'

– JEAN GIRAUDOUX,
THE MADWOMAN OF CHAILLOT, 1943

We return to the question. Do signals from aliens criss-cross the galaxy, aimed from star to candidate star? We know of some signals that do.

It came out of the darkness from the direction of a giant red star. It had been travelling through space for just over twenty years. Reaching the outer edge of the star system, it encountered a disc of rocky and icy debris containing about ten times the number of comets found in our own solar system's Kuiper Belt. These comets spent most of their time frozen in these cold outer reaches, only a few would ever have their orbits perturbed in such a way that they would find themselves approaching their star, becoming hotter and forming a transient tail of gas and dust and perhaps delivering water to the inner worlds. A star like this one would have had 2 billion years or so for a substantial amount of water to be delivered from the outer belts to the inner system and, of course, to any potentially habitable worlds that might reside there. The water flowing from the edge of interstellar space would make all the difference.

The star is not very bright as stars go and very young for

its type. It's a so-called red dwarf, low mass – about a third of our Sun's mass – and low temperature, but as we shall see later it could be an example of one of the most promising places for life to develop in the universe, offering more places and longer timescales than worlds like our Earth and stars like our Sun.

The planets circling this red star have masses between two and fifteen times that of the Earth and all are located within 20 million miles of it, although a Neptune-class world further out is a possibility. Researchers using the Herschel infrared telescope in orbit, which is particularly sensitive to dust and warm gas, have detected a considerable amount of dust which they attribute to cometary collisions, which could be triggered by a planet orbiting near the debris disc. From what we know, the inner worlds of this system might be suitable for life. If their inhabitants detected the signal we sent that flashed through their system and looked closely at it, they would see that it did not in fact come from a red supergiant star but a smaller yellow one in almost the same direction – our Sun. The Gliese 581 complex is the addressee of the interstellar messages: 'A Message From Earth' and 'Hello From Earth'.

But even if there were intelligent aliens in this system and even if they had radio telescopes looking at the right time in the right direction and receiving the right frequency, would they have picked up the signal? Could any aliens out there have picked up any signals from us? Have we given the game away? Do they know we are here?

A decade after the attempts to listen to Mars in 1924, the British Air Ministry authorised the secret construction of five listening stations along the east coast of England, each was to use radio beams in the microwave region of the spectrum to sweep across the sky to warn of any surprise attack by Germany's new Luftwaffe. Later a further fifteen of these stations were installed and, during the war that followed the system became known as radar. Radio transmitters and receivers of ever greater power were developed until, by 1946, it was possible to bounce radar signals off the Moon. The first thing we sent to touch the Moon and the planets were radio waves.

One of the most powerful transmitting dishes was at Millstone Hill in Massachusetts and in February 1958 it was used to send radar pulses towards Venus – our nearest celestial neighbour after the Moon. The output of the radio telescope was scrutinised at the time a reflection from Venus was expected, but it seemed it was all noise, no one was sure if the attempt had worked so when Venus came close to the Earth again the following year another attempt was made, again with equivocal results. Success was achieved at the third attempt in 1961. A power of 12,600 watts was beamed towards Venus, of which, due to the dilution of the signal as it propagated, only about 10 watts reached the planet. Most of which was absorbed and only a fraction reflected back to Earth where a hundredth of a billionth of a billionth of a watt returned to strike the waiting radio dish. However, such is the power of radio astronomy and its ability to detect tiny signals, that this signal was still ten times the level of background noise, which meant it was easily seen. Soon, radar echoes were received from the Sun and other bodies of the solar system. It was an initial demonstration of the principle of sending a message to the stars and astronomers didn't wait long in trying. Project Ozma had listened, now it was time we spoke!

The first message sent into space intended for extraterrestrial civilisations was also used to test the radar radio telescopes ability by once again bouncing the signal off Venus. It consisted of three words, all encoded in Morse code: the word '*Mir*' (meaning both 'peace' and 'world' in Russian) was transmitted from a radio telescope in Yevpatoria, Crimea on 19 November 1962, along with the words 'Lenin' and 'USSR'. Later, these words were beamed towards the multiple star system that contains the red giant HD 131336, which is just over 1,000 light years away in the constellation of Libra (as is Gliese 581). You cannot see this rather undistinguished star with the unaided eye, and although it has been catalogued it has not been intensively studied. We have sent many messages to the nearest stars, at least eleven times, most of them more promising candidates for life than HD 131336. Not all of them were serious attempts to send a signal to aliens, far from it, they were gimmicks and at best they were more of a message to ourselves. Most of the messages were sent from the radio telescope at Yevpatoria in the Crimea.

A more powerful transmission was sent out on 16 November 1974 to celebrate the upgrading of the giant 305-metre diameter Arecibo radio observatory in Puerto Rico, which sadly recently collapsed in spectacular fashion. Built in 1963, utilising a giant sinkhole in the karst topography of the region, its wire-mesh surface was replaced by aluminium panels in the early 1970s. The director of Arecibo was none other than Frank Drake, who with his assistant Jane Allen came up with the idea of beaming a celebratory signal to the stars. Using the technique devised after the Order of the Dolphin conference, they created a message with a total of 1,679 binary digits which, when arranged in a rectangle of 73 rows by 23 columns, revealed a pictorial message.

It contained a lot of information. It began with a lesson in the binary counting system (unlikely to be needed as binary is the simplest possible counting system) and then included the atomic numbers of the five elements that make up DNA – hydrogen, carbon, nitrogen, oxygen and phosphorus – as well as the chemical formula for DNA and a diagram of the DNA helix. There was the population of Earth in 1974 (4.3 billion) and a rough graphic of a stick-figure human. At the base of the pictogram was a diagram of the Sun and the nine planets of the solar system with the third planet – Earth – standing out. The message ended with a graphic of Arecibo itself.

It was sent to the globular star cluster M13 purely because it was in the telescope's line of sight. As the instrument was built into the ground it had very limited pointing ability. This was a demonstration of human technological ability rather than a serious attempt to contact aliens. M13 is a spherical group of ancient red stars about 25,000 light years from Earth on the outskirts of our galaxy. It's among the oldest objects in the universe. It was to be many years before another radio message was beamed towards the stars, and some of these could only be described as bad jokes at best.

In 1999, the RT-70 planetary radar dish at Yevpatoria was again in action, sending two sets of messages to nearby stars: Cosmic Call 1 in 1999 and Cosmic Call 2 in 2003. They repeated the Arecibo message and added a further message in a language called Lincos, which had been devised in 1960 by the mathematician Hans Freudenthal for potential use in communicating with an alien civilisation. It used mathematical rules instead of human syntax. The aim was to send a Lincos 'dictionary' – what the astronomers involved called their 'Interstellar Rosetta Stone' – along with their message. These attempts at communication were a short-lived affair as the Texas-based start-up behind them quickly went out of business. At about the same time there was also the so-called the Teen Age Message, a series of transmissions from Yevpatoria to six Sun-like stars in the Autumn of 2001. The content was chosen by a group of teens from four Russian cities and included a live theremin concert, as well as images and text. In 2008, the song 'Across the Universe' by the Beatles was transmitted by NASA in the direction of the star Polaris by the 70-metre dish near Madrid, which is part of its Deep Space Network. It was to mark the 40th anniversary of the song's recording and the 50th anniversary of NASA. Lone Signal was a crowdfunded project based at the Jamesburg Earth Station in California. On 17 June 2013, it transmitted 144-character messages to the red dwarf star Gliese 526, some 17.6 light years away. Due to a lack of money, transmission ceased soon after it began.

Arecibo was back in action in 2012 on the 35th anniversary of the 'Wow!' signal. This time it transmitted 10,000 Twitter messages towards the stars Hipparcos 34511, 33277 and 43587. A later transmission called 'A Simple Response to an Elemental Message' consisted of 3,775 replies to the question: 'How will our present, environmental interactions shape the future?' A so-called message in a bottle was transmitted in October 2016 by the European Space Agency's Cebreros deep-space tracking station again towards Polaris, approximately 434 light years from Earth, and shortly afterwards 'A Message from Earth' was beamed towards to Gliese 581. Luyten's star, another red dwarf 12.4 light years away, was sent messages in 2017 and 2018. It was at the time known to be home to one planet.

These messages are, to be candid, a motley collection. Nobody has been in charge of sending such messages, and there are no rules other than having access to a suitable equipped radio telescope. There are some limp international protocols about signals from Earth but they are ignored if they are even known at all. If these signals were a serious attempt to represent humanity to the cosmos, then we should as a species bow our heads in embarrassment if not shame. It is therefore good that they are all irrelevant, as we shall see.

Some maintain we are placing ourselves in danger. Jared Diamond, professor of evolutionary biology and Pulitzer Prize winner, said: 'Those astronomers now preparing again to beam radio signals out to hoped-for extraterrestrials are naïve, even dangerous.'

Some say the vastness of space will protect us and that aliens would be far away, too far to be a threat to us. But this is no counter argument at all as we might be 'lucky' and find an alien source much closer. When I was a young scientist at Jodrell Bank, I asked Sir Bernard Lovell about aliens. He had thought about them often, even before he received the now famous letter from Cocconi and Morrison. He replied: 'It's an assumption that they will be friendly – a dangerous assumption.' When I was discussing this issue with the late influential astronomer Zdeněk Kopal, he grabbed me by the arm and said in a serious tone: 'Should we ever hear the space phone ringing, for

God's sake let us not answer. We must avoid attracting attention to ourselves.' The opinion that it is dangerous to assume that aliens are friendly is held by many leading scientists in the field. Physicist Freeman Dyson, formerly of the Institute for Advanced Study in Princeton, said: 'It is unscientific to impute to remote intelligences wisdom and serenity, just as it is to impute to them irrational and murderous impulses.' The Nobel Prize-winning American biologist George Wald took the same view. He said he could think of no nightmare so terrifying as establishing communication with a superior technology in outer space.

Carl Sagan also worried about so-called first contact, advising that we should listen for a long time before taking any action. He added that there is no chance that two galactic civilisations will interact at the same level, one will always dominate the other. There is also the saying that it takes two to be friends, but only one to make war. Others have put it more graphically, saying that the civilisation that blurts out its existence might be like some early hominid descending from the trees and calling, 'Here, kitty!' to a sabre-toothed tiger.

However, Frank Drake is less worried: 'As I thought in 1974, the objections to sending interstellar messages were naïve and carried no weight. The argument then, as now, is that humanity has been, and is making, its presence known through our TV and radio and military radars which, in many cases, release most of their radiated power into interstellar space.'

Science fiction author David Brin has argued that the body ostensibly in charge of drafting policy about sending signals into space – the International Astronomical Union – is being irresponsible. He doesn't want any new transmissions. 'In a fait accompli of staggering potential consequence,' he said, 'we will soon see a dramatic change of state. One in which Earth civilisation may suddenly become many orders of magnitude brighter across the Milky Way – without any of our vaunted deliberative processes having ever been called into play.'

There is one partially accepted protocol concerning contact with aliens, although it is voluntary and toothless. It is called the 'Declaration of Principles Concerning Activities Following the Detection of Extraterrestrial Intelligence'. It is a plan for what should be done if scientists detect an alien signal or some kind of celestial engineering evidence indicative of life and intelligence, such as a probe in our solar system. The plan is mostly about verification and the announcement about contact but it also states: 'No response to a signal or other evidence of extraterrestrial intelligence should be sent until appropriate international consultations have taken place.' I don't think these hopeful words will be followed, and not all scientists think they should.

Michael Michaud, a former US diplomat who was chairman of the Transmissions from Earth Working Group – a division of the International Astronomical Union's Search for Extraterrestrial Intelligence Study Group established in 2001 – almost resigned on several occasions because of the lack of debate about the issue. He believes it is being limited to a narrow, unrepresentative group of scientists with the same attitudes and prejudices and guileless optimism about aliens. He wants the group to be expanded and the topic more widely debated. But this isn't happening, far from it. No one has the authority to stop anyone with suitable equipment transmitting signals from earth or even demand a discussion beforehand. Should we just let our culture be sent into space by groups of people who think they know what's best for us all? When a draft of the protocol was being discussed, the section about deliberate transmissions was deleted, which resulted in resignations from the International Astronomical Union committee in protest and recriminations from both sides. The familiar arguments were wielded about our radio and TV leakage having already revealed our presence and that the great distances involved will protect us. Both of these assumptions are untrue, as we shall see. In any case, any unofficial suggestions will be ignored should contact be made. Most astronomers have never heard of them.

To raise the profile of the debate, if not its tone, opponents to the sending of deliberate signals issued an open letter in 2015. One of its signatories was John Gertz, who said that, as far as he was concerned, such signals were equivalent to unauthorised diplomacy. Scientist Doug Vakoch, clearly frustrated by the fact that we have never found any evidence of aliens, was in favour of signals and he promptly left his position at the SETI Institute in California to found Messaging Extraterrestrial Intelligence (METI) International to promote his desire for messaging. 'If I had my way,' he said, 'when Frank Drake did his first SETI search in 1960, he would have also begun transmitting, and if he had we would be looking for responses from civilisations out to a distance of 30 light years by now.'

One thing is clear from our searches for extraterrestrials – there is nobody transmitting strong interstellar beacons in our local vicinity. If 'they' are out there, they are keeping quiet, prompting the realisation that they might know something we don't.

Despite what science fiction shows us, interstellar travel will be far from easy, if it is really possible at all. *Star Trek* and *Star Wars* would not be the same if it took years or decades to travel from star to star. Travelling at the speed of light, it would take 100,000 years to traverse the Milky Way, and we have only the vaguest ideas about faster than light travel, which is probably practically impossible even if theoretically possible in some extreme circumstances.

We, and the aliens, will have to make do with slower-than-light travel, and there are many possibilities. Nuclear fusion power could propel a starship but probably only at about 10 per cent of lightspeed. This is not nearly fast enough for *Star Trek* but a capability that could colonise the galaxy given enough time. Perhaps after several long-duration star-hops out into the galaxy, interstellar travellers would forget where they came from like the citizens of the Galactic Empire in Isaac Asimov's *Foundation* trilogy, who debate about the discredited theory that Earth may have been the home world of humans.

We are young, a 100-year-old technological civilisation living in a 10-billion-year-old galaxy. To older civilisations we might be a minor phenomenon, perhaps not even properly intelligent by their estimation. Only those civilisations close to our own technological level may find us interesting, we don't know. Perhaps they or older cultures might leave aliens like us alone and allow us and others to develop in our own way. One can think of many such possibilities and scenarios, but they all remain assumptions and we have no idea if they are reasonable or not. Charles Fort, the American writer and researcher who specialised in strange phenomena, wondered if we would be interesting enough: 'Would we, if we could, educate and sophisticate pigs, geese, cattle? Would we establish diplomatic relations with the hen that is satisfied with its sense of achievement?'

Aliens could have had to become brutal to survive and not wise and peaceful as many naïvely wish. The Australian astronomer Ronald Bracewell warned that other species could place an emphasis on cunning and weaponry, as we often do, and that an alien ship dispatched our way would be likely to be armed. Indeed, evolution on Earth is, as they say, red in tooth and claw. It is conceivable that any aliens we contact will also have had to fight their way up their evolutionary ladder and may possibly be every bit as nasty as we can be – or worse. Imagine an extremely adaptable, extremely aggressive super-predator with superior technology that is able to traverse interstellar space and destroy worlds.

Visits from alien spacecraft are a popular subject in the media. Steven Spielberg's films *Close Encounters of the Third Kind* and *E.T. the Extra-Terrestrial* portrayed aliens as enigmatic and friendly, inspiring even. E.T. was harmless. Films such as *Independence Day* show alien visitors as absolutely ruthless and the 1951 film *The Day the Earth Stood Still* featured a sinister humanoid alien sent to warn humanity about its warlike tendencies. *The Thing from Another World*, also from 1951, depicted a creature who wanted to wipe out us humans and replace us with its offspring.

But in 1992, Frank Drake wrote that we should not fear contact with aliens. Unlike the fate of primitive civilisations on Earth that were overwhelmed by more advanced societies, humanity could not be harmed because the aliens would be too far away. SETI pioneer Bernard Oliver said interstellar travel was impossible and so there was no reason to remain quiet. However, Carl Sagan has written that radar and television leakages from a newly emerging technical society such as ourselves could induce a rapid response by nearby aliens who want to reach us: 'If interstellar spaceflight by advanced technical civilizations is commonplace, we may expect an emissary, perhaps in the next several hundred years.' A report to the US Congress on the contact with extraterrestrials said: 'The receiving civilisation might be capable of interstellar flight and dispatch emissaries for further investigation. With no foreknowledge of their character, we might be aiding in our own doom.'

Recent calculations indicate that any human transmissions into space will fade far more quickly than we thought, so we may be safe after all.

Contact would start with a mysterious object emerging from the dark heading towards us.

Such an object was detected on 19 October 2017 by the Pan-STARRS1 sky survey telescope situated high on an extinct volcanic Hawaiian island.

EMISSARY

An observatory on an extinct volcanic Hawaiian island, above a third of the Earth's turbulent atmosphere, is an ideal place to scan the sky. The object was also seen by the Canada–France–Hawaii telescope on nearby Mauna Kea. Its swift motion stood out from the other objects – comets, asteroids, variable stars. It was immediately obvious that this object was unique, travelling too fast to be part of our Sun's family. It was a visitor from beyond. It came from deep space and nothing could prevent it from returning. It was the first interstellar visitor ever detected.

It came into our solar system travelling at great speed, passing the orbits of the giant planets unnoticed as it fell sunward. Its surface was tarnished and dulled by hundreds of millions of years of exposure to cosmic rays as it traversed the vastness of interstellar space. Yet something could still be made out, markings on its side, emblems of some sort? Perhaps its wrinkled micro-thin sail moved slowly when tugged gently by one of its few remaining tethers. The craft was tumbling chaotically. Inert for aeons, perhaps it carried a dead cargo. It gathered speed as it got closer to the yellow dwarf star it was rounding. It would not stay in this planetary system for long.

Excitedly, observers from all over the world turned towards it. The so called Very Large Telescope in the Atacama Desert of northern Chile, the Gemini South Telescope also in Chile and the Keck II also on Mauna Kea took data. Soon, the Hubble

Space Telescope and the Spitzer Space Telescope were on its trail. Astronomers looked through previous observations to see if it had been seen before its discovery. It was obvious it was small and not very luminous. Spectra obtained on 25 October showed it to be reddish in hue, resembling the spent nucleus of a comet. Later that day, the William Herschel Telescope in the Canary Islands detected no details in its spectrum. It seemed to be similar to our asteroids or comets, albeit from far beyond our solar system. But then astronomers noticed something peculiar about how the object was dramatically changing in brightness.

It was tumbling in space and from the way its brightness fluctuated it was not roughly spherical as was to be expected but elongated like a giant cigar with a length to width ratio of 1 to 5 or maybe greater. NASA's Jet Propulsion Laboratory issued a news report saying it was 400 metres long and ten times longer than it was wide. Was it a spaceship? The internet went wild, and then the object deviated slightly from its predicted trajectory. Was it changing course?

The Green Bank radio telescope in West Virginia looked at it for six hours to see if it had any radio emissions that indicated it could be artificial. Nothing was heard. One astronomer said if there had been a cell phone on its surface, they would have picked it up. Oumuamua, as it was called, is Hawaiian for the arrival of the first distant traveller. It caused a sensation.

It rounded the Sun on 9 September 2019 and on the outward leg of its journey it passed beyond the orbit of Earth on 14 October at a distance of 24.2 million kilometres. It reached the orbit of Mars on 1 November, later passing the orbits of Jupiter and Saturn, slowing down as it climbed out of our Sun's gravity and returning to the same speed as it had before its approach to our solar system. It is now heading towards the constellation of Pegasus having swung 66 degrees from the direction of its approach.

Was this object a spaceship from the stars? The first extraterrestrial emissary? Only a few think this is a possibility. The evidence for it being a spaceship is slim to say the least. It is a natural object, but it is one that focused minds. If a chunk of rock can travel between the stars, then why not a spaceship? And what is so special about today? It's more likely to have occurred in the past, many times.

Sometime in the next few decades, we will transform the exploration of our solar system by using small multi-tasking probes directed by artificial intelligence. We will send swarms of them to fly and crawl over the surface of Mars, to drift through the clouds of Saturn's major moon Titan and to swim in its hydrocarbon seas. They will descend the fissures of Saturn's other moon Enceladus, with cracks in its icy shell allowing access to its sub-surface ocean of warm water and the enticing prospect of life. They will also move out into the asteroid belt, visiting, surveying and cataloguing these various chunks of rock, seeking out those with exceptional metals, ice and mineral content for further study and possible mining.

One of those probes might find it.

Might this be one of a colonising wave of self-replicating probes that move through the galaxy? They've been called von Neumann probes, after the American physicist John von Neumann, who in 1948 described a self-replicating machine. Once the daughter probe is built from raw materials found in a solar system asteroid belt, it can go off and build more probes and so on. Twenty years later, astronomer Ronald Bracewell envisaged swarms of such probes gathering information and compiling a galactic database. Alternatively, science fiction writers have imagined probes designed to eliminate any intelligent life that they find.

Perhaps its alien nature will be apparent immediately, its unnatural shape or radiation signature. What is its purpose, and is it still functioning? Perhaps its builders sent it to explore, along with many others, to endure the millions of years traversing between the stars. Were they scattered among the stars like dandelion seeds or targeted at our own solar system? Perhaps it was sent by a civilisation living in a neighbouring star system specifically towards us, because they were aware of our leakage or our deliberate transmissions, to observe and report on us. Perhaps it is from much further away in space and time? Could the ancestors of this artefact have set sail into our galaxy hundreds of millions of years ago, moving between planetary systems using the resources it might find there to build copies of itself to stay behind or undertake further expansion across the galaxy?

The maths of such an enterprise is straightforward yet staggering in its implication. It suggests that a colonising wavefront could spread across the galaxy within 700,000 years, so aliens, robotic or biological, have had plenty of time to colonise our galaxy, so where are they? If the above line of reasoning is correct, then alien probes should be everywhere. So, is there one here somewhere in our solar system? How would we know?

It has been suggested that a good place to hide would be the asteroid belt as there are a great many objects among which to mingle as well as resources from which to build copies. For monitoring the Earth, a solar orbit would be better, closer and not easily detectable. Another good strategy would be to position itself at a so-called Lagrange gravitational balance point at which it's possible to maintain a surveillance position with little expenditure of energy. There are several of these points in relation to the Earth. We have placed many space observatories in these regions, and they haven't bumped into any alien probes, we've looked.

Is it just an artefact? Something that came here and died, failed in some manner or was discarded? Years after its discovery by our AI probe, we would send a more capable probe or more likely a crewed mission to conduct a delicate task of space archaeology. What tales, I wonder, could a dead alien probe tell? Would it contain the alien equivalent to our Voyager golden disc – our brief summary of humanity? Would it house something grander? An encyclopaedia or a history of an alien race. Would we awaken an avatar to converse with, or would we begin the task of picking apart inert alien tech? Perhaps these sentinels are closer to us than the asteroid belt?

In his famous story of 1951 *Sentinel of Eternity*, the science fiction writer Arthur C. Clarke envisaged an object left on the Moon. It caught the eye of an astronaut undertaking a tractor survey. Something glinting from atop a mountain, something that should not be there. This sentinel was then used as the starting point for the 1968 novel and film *2001: A Space Odyssey*. The narrator of the story speculates that the tetrahedral structure made of a polished material may belong 'to a technology that lies beyond our horizons'. In Clarke's story, which despite being initially unsuccessful went on to transform his career, the sentinel has been transmitting information for millions of years, presumably reporting on the development of life on Earth and especially the rise of humans. I wonder what it said when it detected the first artificial radio waves, and the first intense flash of gamma rays indicating we had acquired the ability to build atomic weapons, the arrival of humans on the Moon and its own discovery.

We have very detailed maps of the Moon that can easily detect artefacts on its surface, though all of them have been left by humans. In all the exquisite detail relayed back by probes such as the American Lunar Reconnaissance Orbiter and Japan's Kaguya, we have found nothing untoward, no structures, cities or markings. If there is a Sentinel of Eternity on the Moon we have yet to find it. But is the same true of Mars? The ideas around astro-archaeology are fascinating, but many who should take a serious view of it are reticent to do so because the subject has a poor reputation. I am thinking about the pyramids of Elysium on Mars.

In 1972, the Mariner 9 spacecraft in orbit of Mars saw features on the plains of Elysium that resembled pyramids. More provocatively, a few years later, one of the Viking orbiters looked at a butte in the region known as Cydonia that under certain lighting conditions resembled a huge human face staring up at the sky. Was this a monument left behind by some long extinct civilisation? Was Percival Lowell right after all about life on Mars? When the Mars Global Surveyor looked again at the

Cydonia region in 1998, the face was gone and was nothing more than a jumble of rocks and ridges. As humans, we are programmed to see faces in patterns from the moment of our birth. There are still some enthusiasts who cling to the idea that there is a face on Mars, along with images that they say show evidence of buildings and cities on the planet, but they are deluding themselves. Nonetheless, the observations made by the Mars Global Surveyor in that particular region of Cydonia was the first time NASA programmed a planetary probe to search for possible evidence of an alien civilisation.

In 1963, Carl Sagan suggested that extraterrestrial civilisations would visit planets with intelligent life every 10,000 years and those with advanced civilisations every 1,000 years. Consequently, he argued, there is a possibility that the Earth was visited at least once in the past. Is there any evidence for this? Scientists, he said, should examine ancient myths and legends for any evidence of contact while being aware of the problems of interpreting another civilisation from another time. He talked of a plausible contact myth in some of the stories surrounding the origins of the Sumerian civilisation in the fourth millennium BC. Archaeologists should also be aware of the possibility that they might uncover extraterrestrial artefacts. This has been called paleo-contact. Swiss author Erich von Däniken has claimed that there have been extraterrestrial visitors to Earth. In his very successful book *Chariots of the Gods?* published in 1968, he said we should take stories about gods who came from on high and walked the Earth at face value. None of it stands up to close scrutiny. But why are these ideas so popular? Von Däniken's latest book was published in 2021. Others have also suggested that alien visitors influenced our past by giving us knowledge that accelerated our development. I can't help but think that such theories reflect a low opinion of human abilities.

We are already an interstellar species as we have sent several objects out of our solar system. Perhaps they will be found by someone or something else. Pioneer 10 and 11 are carrying a graphic message in the event an intelligent life form may capture either spacecraft on its journey. Engraved on a gold-anodised aluminium plaque, the message features a drawing of a man and a woman, a diagram of our solar system and a map depicting our galactic position with reference to objects known as pulsars. The man has his hand raised in a friendly welcoming gesture.

Voyager 1 and 2 contain a much more sophisticated message contained on the golden discs we mentioned earlier. They both have one on their sides that contains 115 images as well as the sounds, languages and music of Earth. There is the Taj Mahal, the Sydney Opera House, DNA, trees, the Great Wall of China and the Great Barrier Reef. There is Akkadian, a language spoken in Sumer 6,000 years ago and Wu, a modern Chinese dialect. There is aboriginal music, the music of the first humans who had enough to eat, Bach, Mozart and Beethoven are there and so is Chuck Berry. There is the sound of the wind and the rain, lightning, a mother singing a lullaby to her child, laughter and a man and a woman kissing.

There is also a written message from US President Jimmy Carter, which reads: 'This is a present from a small, distant world, a token of our sounds, our science, our images, our music, our thoughts and our feelings. We are attempting to survive our time so we may live into yours. We hope someday, having solved the problems we face, to join in a community of galactic civilizations. This record represents our hope and our determination, and our goal in a vast and awesome universe.'

In February 1990, Voyager 1's cameras were turned on for one last series of photos, a 'family portrait' showing seven of the solar system's nine known planets in an unprecedented mosaic. Voyager 1's cameras had been inactive since it left Saturn ten years previously. Voyager 1 was 6 billion kilometres distant, so only Jupiter showed as a discernible disc. Our Earth is just a point of light. High above the plane of the planets, Voyager has a unique view from the outside. One scientist said that the pictures symbolised our place in the cosmos, then he looked away rather sadly. These will be Voyager's last pictures.

We will lose contact with Voyager 1 and 2 as they begin their drift among the stars. Aeons hence, long after the Earth has been destroyed by our dying Sun and perhaps long after man is extinct, those whispers from Earth will still be floating among the stars. They could be our last mark on the cosmos. The Voyager crafts and their messages could be what we are judged by.

In Arthur C. Clarke's novel *Rendezvous with Rama*, a giant alien spacecraft passes through our solar system. Astronauts land on it and explore its interior but it goes on its way without any communication. Modern survey telescopes can scan the sky several times a night looking for things that change and move. Perhaps sometime, tonight possibly, something will be seen that on second viewing seems out of the ordinary.

HELLO EARTH

In October 1938, a radio adaptation of H.G. Wells' *The War of the Worlds* caused consternation in the United States, much to the surprise of those behind the radio play who were unimpressed by the script. When it was recorded, the narrator and lead actor of the Mercury Theatre of the Air, Orson Welles, said it was 'corny' and cursed the writers, saying the whole show would be silly. Little did he know.

The novel aspect of the broadcast was that the scriptwriter Howard Koch decided to change the setting of the site of the invasion in the story from Woking in England to America, selecting the New Jersey village of Grover's Mill. He used news bulletins interrupting a programme of music being broadcast from a hotel. Frank Readick played Carl Phillips, the reporter on the scene who described the invasion before collapsing dead at his microphone. Readick had listened to a recording of the report of the explosion of the *Hindenburg* air balloon and based his performance on the commentator's horror at the catastrophe.

It began with: 'CBS presents Orson Welles and the Mercury Theatre on the Air in a radio play by Howard Koch suggested by the H.G. Wells novel *The War of the Worlds*.' No one at the Mercury Theatre ever thought that anyone would think an actual invasion from Mars was taking place.

On this particular evening, 30 October, it was estimated that 32 million people were listening to the radio, although not many were tuned in at the start of the Mercury broadcast.

It was at 8.12pm that things changed. Most had been listening to the popular *Edgar Bergen and Charlie McCarthy Show*, which improbably featured a ventriloquist and his unruly dummy. At 8.12pm there was an intermission with a singer and a large proportion of the listeners reached for their dials, intending to return to the dummy after a few minutes. However, that night they stumbled upon a news report of an invasion, by now well under way, by Martians. It didn't sound like a spoof.

> Ladies and gentlemen, I have just been handed a message that came in from Grover's Mill by telephone. Just a moment. At least forty people, including six state troopers, lie dead in a field east of the village of Grover's Mill, their bodies burned and distorted beyond all recognition.

Enter Carl Phillips:

> Good heavens, something's wriggling out of the shadow like a gray snake. Now it's another one and another. They look like tentacles to me. There, I can see the thing's body. It's large as a bear and it glistens like black leather. But that face, it … Ladies and gentlemen, it's indescribable. I can hardly force myself to keep looking at it. The eyes are black and gleam like a serpent. The mouth is V-shaped with saliva dripping from its rimless lips that seem to quiver and pulsate … This is the most extraordinary experience. I can't find words … I'll pull this microphone with me as I talk … Hold on, will you please, I'll be right back in a minute.

This was followed by the sound of a microphone hitting the ground, and then silence. The heads of the US armed forces were brought to the microphone, with the secretary of the interior saying: 'We must continue the performance of our duties each and every one of us, so that we may confront this destructive adversary with a nation united, courageous, and consecrated to the preservation of human supremacy on this earth.'

It is unsurprising that a small but not insignificant proportion of the audience, mostly in the New Jersey area, were somewhat alarmed and the Mercury audience had doubled to 6 million. The CBS switchboard was jammed. One operator told the callers: 'I'm sorry, we don't have that information here.' Three times during rest of the programme, listeners were reminded that they were tuned in to the Mercury Theatre of the Air performing an adaptation of H.G. Wells' *The War of the Worlds*. But some listeners had already made their minds up.

Some tried to call relatives but found the lines were engaged. Some took to the streets; some went to church. In Harlem, a congregation fell to its knees; in Indianapolis a woman ran screaming into a church: 'New York has been destroyed. It's the end of the world. Go home and prepare to die.' In Newark, New Jersey, occupants of a block of flats left with wet towels around their heads as improvised gas masks. In Staten Island, Connie Casamassina was just about to get married, when they heard the news she started to cry, sobbing, 'Please don't spoil my wedding day.' The Mienerts family of Manasquan Park, New Jersey, along with their dog, sped down the street in their car, puzzled that passers-by knew nothing about the invasion.

The panic even reached the studio with a shaken Welles announcing that the programme had 'no further significance than as the holiday offering it was intended to be. The Mercury Theatre's own radio version of dressing up in a sheet and jumping out of a bush and saying "Boo!" ... So goodbye, everybody, and remember, please, for the next day or so, the terrible lesson you learned tonight. That grinning, glowing, globular invader of your living room is an inhabitant of the pumpkin patch, and if your doorbell rings and there's no one there, that was no Martian ... it's Halloween.'

One of the producers of the show later said: 'Someone had called threatening to blow up the CBS building, so we called the police and hid in the ladies' room on the studio floor.' Reporters besieged the building asking how Welles felt about the many deaths the broadcast had caused. Frightened CBS put out hourly disclaimers. By that time, up in lights in Times Square was the moving news sign: 'ORSON WELLES FRIGHTENS THE NATION.' The following day, newspapers devoted much space to descriptions of what had happened and the gullibility of the American people in the face of the 'irresponsible' power of radio, at a time when newspapers had been losing the battle with radio as the chief disseminator of news.

Welles was initially shocked but later reflected: 'The most terrifying thing is suddenly becoming aware that you are not alone. In this case the Earth, thinking itself alone, suddenly became aware that another planet was prowling around.' He then added: 'The last two generations are softened up because they were deprived in their childhoods, through mistaken theories of education, of the tales of blood and horror which used to be part of the routine training of the young. Under the old system the child felt at home among ghosts and goblins, and did not grow up to be a push-over for sensational canards. But the ban on gruesome fairy tales, terrifying nurse-maids and other standard sources of horror has left most of the population without any protection against fee-fi-fo-fum stuff.'

It could be said that the publicity Welles got took him to Hollywood where he made *Citizen Kane*, but he seemed annoyed that throughout his life he was always being asked about the radio broadcast, and later rewrote history to suit himself and falsely claimed that he wrote the script: 'Now it's been pointed out that various flying saucer scares all over the world have taken place since that broadcast … everyone doesn't laugh any more. But most people do. And there's a theory this is my doing. That my job was to soften you up … ladies and gentlemen, go on laughing. You'll be happier that way. Stay happy as long as you can. And until the day when our new masters choose to announce that the conquest of the earth is completed, I remain, as always, obediently yours.'

SETI scientists do not want their objective linked to spaceships, invaders from Mars or UFOs as they believe such things threaten their credibility. Jill Tarter of the SETI Institute in California once wrote a letter to a newspaper that had a story about searching for life in space positioned alongside a feature about UFOs. She said that real science should be distanced from pseudo-science and that irresponsible journalism could jeopardise government funding for her subject. Not too long afterwards the US government did indeed cancel all of its funding for SETI.

The fact remains though that the most popular and controversial image of contact with aliens involves objects that some believe are vehicles from other worlds. For most people, there are two ways they encounter the subject of life in space: science fiction and the UFO phenomenon along with its stories of alien contact and abduction. Over the years, tens of thousands of people have reported seeing unexplained objects in our skies and this is probably an underestimate. In the early 1950s, the US Air Force estimated that only one in ten people who had seen UFOs reported their experience, and they were not stupid people. They included scientists, air-traffic controllers, pilots, police officers and military personnel – credible people.

We must be careful here. One report stated that 2 per cent of Americans stated they had been abducted by aliens, but looking at such reports a little deeper shows that things are not quite that straightforward. In this case, the researchers did not ask direct questions but meandered around the topic with questions like: 'Have you ever woken up paralysed with the sense of some strange presence in the room?' This is not explicit enough. If respondents answered enough of the somewhat vague questions they were classified as having had an alien contact.

The media in all its manifestations loves aliens, they are everywhere and there is an established folklore about alien sightings and abductions. I don't believe any of it, and I will tell you why.

I love the aliens in the bar scene in *Star Wars: A New Hope*, a sleazy dive in the spaceport of Mos Eisley. Some of the strange creatures in the scene are playing odd instruments, but the aliens themselves are not strange at all, they are all slight variations of humans, in fact they represent cultural icons we are all familiar with. There are smooth snake-like creatures and furry, fox-like ones, almost straight out of Br'er Rabbit. Our 'modern' aliens are stand-ins for fairies, elves and ghosts; the vampires and spirits of the meres and dark woods. Their appearance taps into supernatural fears, cultural echoes of the time when we looked at the dark with dread and superstition and saw incubi, succubi and witches, but now they are rendered harmless, Orson Welles was right.

What about those who claim to have been interrupted by aliens? The creatures they claim to have seen are, like those in the bar at Mos Eisley, far too anthropomorphic. They have faces almost like ours and similar body plans, which reveal that their far ancestors must have been Earth fish (a horizontal mouth below the eyes and nose). It is amazing that these beings from another star are so similar to terrestrial land vertebrates. It is also curious that the stories people tell about alien encounters are remarkably consistent. They are woken in the dead of night or taken from their car or workplace by large-headed, small, thin-bodied, huge-eyed, grey aliens, who take them to their spacecraft, which is usually silver and gleaming with sparse corridors, and they are often subjected to medical and gynaecological procedures. Then they wake up in bed. Such experiences are imaginings, what is commonly termed 'waking dreams', or sleep paralysis, when our brains are not in sync with our bodies in the transitions through sleep periods. When people are affected this way, they already know the plot. Some people claim to have had implants inserted into them, but analysis of such artefacts has been disappointing.

However, there is more to UFOs than my outright dismissal indicates, and they can teach us things about life in space. There are strange things that have been seen in the sky that defy explanation, and the US government recently even admitted to this after a report given to them in the spring of 2021. Many surveys show that the public are happy to believe in intelligent life on other worlds and that there is the possibility they could be visiting our planet. The UFO phenomenon emerged from the public, and it was scientists who 'generated the emotional storm against allowing unprejudiced examination of observations made by persons judged sane by conventional standards', as claimed J. Allen Hynek, the astronomer who was the adviser on the movie *Close Encounters of the Third Kind*. He added: 'The UFO evidence has not been properly presented at the Court of Science.'

There is an obvious contradiction here and one that the public at least is aware of. Frank Drake has said there is no evidence that any UFO is the product of aliens and has gone on to argue that if UFO reports are real, they must be due to extraterrestrial spacecraft, but as interstellar travel is impossible, UFOs should be dismissed, so there is no phenomenon to study. You see the circular self-serving distortion of logic there. Indeed, many SETI scientists do not even consider any form of direct contact with aliens – UFO, alien emissary etc. – to be worth studying.

Many years ago, I was briefly on the SETI Committee of the International Astronomical Union and I was delighted to be asked to join and be party to the discussions. Unfortunately, my stay was brief and I wasn't able to attend their meetings held all around the word because of my journalistic day job at the

BBC, unlike other committee members who were at institutes and academic institutions. Nonetheless, I was part of heated discussions about what was called 'Protocol 3', which concerned drawing up a policy of what to do if an alien spaceship arrived on Earth. Opinions were split, some thought having a protocol was a wise move, others vehemently argued that it was playing into the hands of the UFO fanatics and legitimised their views, tainting the serious consideration of the search for intelligent life in space. Protocol 3 was not adopted.

My view is that there is a phenomenon to study and also that contact through interstellar probes is at least as credible as receiving radio messages from the stars, even if the two are unconnected. Do UFOs have anything to do with interstellar probes? If there are no aliens, then what have people seen? And even if there are no aliens, does what has happened, is happening, tell us something about ourselves and our deep-seated, primal attitudes to life in space?

This is not a new phenomenon. Reports of flying objects with alien or humanlike occupants go back to antiquity, where they are intertwined with religion, myth and superstition. A Chinese tale relates a distant land of flying carts flown by one-armed, three-eyed people. Ancient Egyptians recorded sightings in the sky. The Roman historian Livy reported strange phantom craft in the sky. Perhaps the most famous of all was in the Old Testament's Ezekiel 1, a written account from around 590 BC: 'A stormy wind came out of the north, and a great cloud, with fire flashing forth continually, and a bright light around it … in the midst of the fire something like glowing metal. And in the midst of it were figures resembling four living beings … they had human form. Each of them had four faces and four wings.' Some interpret this as an actual aerial craft, while others see it as a supernatural experience or a hallucination. There have been other accounts. In Japan in 1235, a warlord observed lights circling in the night sky. He asked wise men what it was, and they said the wind was making the stars sway.

People are not stupid and less easily fooled than many would like to believe. Sometimes, individuals insist they have seen something and stick to their stories despite public denial and disbelief. And sometimes they are right.

The first relatively modern, well-recorded series of sightings of strange objects in the sky in the United States occurred in late 1896. Thousands of people spread over many states said they had seen airships and talked to their crews. Many commentators at the time dismissed the sightings as imaginary, saying that it was to be expected that the first flying machines would be airships and that people were seeing what they expected to see. Some said the airships they had seen were spacecraft from Mars.

Those who dismissed the sightings, or thought they were aliens, were wrong. A dirigible-like balloon had flown over Paris in 1852, and one flew in the United States in 1865. Researcher Michael Busby studied hundreds of press reports, mapped the sightings and wrote a book, *Solving the 1897 Airship Mystery*. It seems that airships had been flying as early as the 1840s and were perhaps even sponsored secretly by the US government. But the sightings did not cause panic or anxiety, because this was a time of optimism for how science and engineering could improve society. When the Cold War came along, that attitude would change dramatically.

Charles Hoy Fort was an influential American writer and researcher who specialised in reporting strange phenomena. His influence persists to this day, as the terms 'Fortean' and 'Forteana' are sometimes used to describe such phenomena. Fort's admirers call themselves 'Forteans'. For more than thirty years, Fort visited libraries in New York and London, collecting notes on phenomena that were not explained by the accepted theories and beliefs of the time. For instance, he suggested that there is a 'Super-Sargasso Sea' into which all lost things go. He claimed that his ideas fit the data and the conventional explanations that were offered. As to whether Fort believed in what he collected, he wrote: 'I believe nothing of my own that I have ever written.'

But there is a point behind such contrariness. The writer Colin Wilson said that Fort's work gave him 'the feeling that no matter how honest scientists think they are, they are still influenced by various unconscious assumptions that prevent them from attaining true objectivity. Expressed in a sentence, Fort's principle goes something like this: People with a psychological need to believe in marvels are no more prejudiced and gullible than people with a psychological need not to believe in marvels.'

Fort published reports of peculiar sightings, some of which would now be categorised as UFOs, and wrote in 1919 that 'our data have been damned, upon no consideration for individual merits or demerits, but in conformity with a general attempt to hold out for isolation of this earth'. He said that the notion of things dropping in on our planet from 'externality' was unsettling to science; the scientific attitude toward the unwelcome is that it does not exist.

Still, such objects came. In the Second World War, some military pilots described glowing objects flying beside their aircraft, although these 'foo fighters' did not become widely known until much later.

The modern era of UFO sightings began in 1947 when pilot Kenneth Arnold described disc-shaped objects flying in formation, making motions like a saucer skipping over water. Newspapers used the term 'flying saucer', with a tone of ridicule, but what this actually did was to provide a frame of reference for people to place such inexplicable observations. People from all over the United States came forward with their own tales of strange objects in the sky and most assumed that UFOs were real but easily explained as something other than alien craft. A 1947 Gallup poll concluded that most people thought they were illusions, hoaxes, secret weapons or other explainable phenomena; few thought they came from outer space.

Swiss psychologist Carl Jung interpreted the UFO phenomenon in the context of the Cold War, seeing them as a manifestation of a hope for a better world: 'Just at the moment when the eyes of mankind are turned towards the heavens, partly on account of their fantasies about possible spaceships, and partly … because their earthly existence is threatened, unconscious contents have projected themselves on these inexplicable heavenly phenomena and given them a significance they in no way deserve.'

It is often difficult to isolate what exactly the UFO phenomenon is, given all the hoaxes and bad investigations, as well as the myths that have developed around the subject. Consider the alleged UFO crash near Roswell, New Mexico, in 1947.

Some claimed that the wreckage included the bodies of aliens that were taken to the now famous Area 51, a US Air Force base at Groom Lake, Nevada, which was established in the 1950s to test the U-2 spy plane. A 1997 Air Force report stating that there were no captured aliens has not stopped the speculation, in fact, quite the reverse.

The US Air Force instigated Project Sign in December 1947 for the study of UFOs (it later became Project Grudge, then Project Blue Book), but it didn't go the way it was planned. Later accounts suggested that Air Force investigators were favouring the extraterrestrial hypothesis while publicly dismissing saucers as natural phenomena or hoaxes. That position was soon to change.

During the Cold War, military control of the investigation of UFOs seemed natural, and if any in the scientific community were interested, they were discouraged. In the climate of secrecy, conspiracy theories proliferated and the government became worried. So-called contactees – people who claimed that they had met aliens and had even been inside their spaceships – started coming forward. According to them, the aliens looked like humans and came from planets free from war, poverty or unhappiness. They wanted to help us prevent war, stop nuclear testing and help us build the kind of utopian society that they enjoyed. In 1953, the CIA recommended that national security agencies strip UFOs of their aura of mystery, and soon astronomer Donald Menzel became the first American scientist to publish a book on UFOs, saying the idea that they were the product of extraterrestrial intelligence was ludicrous. People who believed it, he said, were lunatics, frightened and confused.

Harvard psychologist Susan Clancy said that abduction stories may give people a deep sense that they are not alone in the universe and that their memories resembled transcendent religious visions – scary and yet somehow comforting.

For years, there was an impasse with the US Air Force on one side and special interest groups on the other. The public were fascinated and books on the subject became bestsellers. Then, in 1965, the press, the public, the US government and the scientific community were all involved in the controversy when the US Congress held the first open hearing on the subject. J. Allen Hynek's provided a testimony that criticised the military and called for a panel of civilian scientists to determine if a major problem actually existed. The University of Colorado asked the physicist Dr Edward Condon to write a report, which was published in January 1969. It produced what many scientists wanted when it said it had found no direct, convincing evidence for the claim that any UFOs were alien spacecraft. It stated that interstellar travel was impossible. Condon acknowledged that people who reported UFOs were normal and responsible civilians and that some sightings were difficult to explain by conventional explanations.

With hindsight, it's clear that the Condon Report was confused. The issue was not just a question of whether UFOs were spaceships, it was also a question of whether there was anything to study, whatever UFOs were. Sadly, nobody saw this problem at the time. *The New York Times* said the report was courageous because it faced up to what was becoming a growing religion. *The Nation* agreed with Condon's recommendation that it was necessary to prevent school children from reading about UFOs as it would give them a warped view of the way science was done.

But whatever the UFO phenomenon was, it was telling us something important about the human reaction to life in the universe. The American Institute of Aeronautics and Astronautics issued a study challenging the Condon Report's conclusion, saying they could not ignore the small amount of well-documented but unexplained sightings. But the US Air Force wanted to close the issue and, in December 1969, the secretary of the Air Force announced it was ceasing all studies of UFOs.

Contradicting Frank Drake, in 1968, Carl Sagan said: 'I do not think the evidence is at all persuasive that UFOs are of intelligent extraterrestrial origin, nor do I think the evidence is convincing that no UFOs are of intelligent extraterrestrial origin.' He concluded that there were what he called 'resonances', involving religion, boredom, novelty, military secrecy and the intolerance of ambiguity. But as I have said, there was another resonance – the desire to be part of something larger.

During his presidential campaign in 1976, Jimmy Carter revealed that he had seen a UFO in 1969. He said that if he became president he would make public every piece of information that the United States held about UFOs. It was a canny move but it did not contribute to his success. After Carter was elected, White House science adviser Frank Press recommended that NASA form a panel to investigate if there had been any new findings since the Condon Report. NASA administrator Robert Frosch was not interested.

Hynek's book *The UFO Experience* is fifty years old now. He did not believe the alien explanation but maintained there existed something worthy of study. 'When the long awaited solution to the UFO problem comes,' he wrote, 'I believe that it will prove to be not merely the next small step in the march of science but a mighty and totally unexpected quantum jump.' He added: 'Nothing that intrigues the mind of Man is automatically ineligible for the scientific approach.' Arthur C. Clarke acknowledged that phenomena still unknown to science may account for the very few UFOs that are both genuine and unexplained.

Hynek appears in the film *Close Encounters of the Third Kind*. He is seen at the end of the film, after the aliens emerge, bearded and with a pipe in his mouth. Astronomer Jacques Vallée was the inspiration for Claude Lacombe, the researcher played by François Truffaut in the film. He said that he tried to interest Spielberg in an alternative explanation for the UFO phenomenon: 'I argued with him that the subject was even more interesting if it wasn't extraterrestrials. If it was real, physical, but not E.T.' Spielberg replied: 'You're probably right, but that's not what the public is expecting – this is Hollywood and I want to give people something that's close to what they expect.'

The UFO phenomenon is relevant to the search for life in space because of what it tells us about ourselves. Writer Randall Fitzgerald said that if we peer through the looking glass of unidentified airships, mutilated cattle and crop circles we see reflected back at us the wondrous dark side of the collective hopes and fears for the future of our species.

In July 2021, the US Department of Defense presented a report to Congress about UFOs, or to use the department's preferred term, Unidentified Aerial Phenomena (UAP). The report was requested by Congress some months before and widely speculated upon by the media after film footage from the cockpits of fighter aircraft had been publicly released showing what were fuzzy blobs moving in strange ways in the sky. The report that was published, which was a summary of a much larger one that might or might not be made public in the future, said there were many instances of UAPs that could not be explained, but that the Department of Defense was sure they were not alien spacecraft. The intelligence report is noncommittal and ambiguous and seemed to satisfy both sides of the debate.

What came out of it is what we have always known: we want to believe.

US AND THEM

'But I've often wondered, what if all of us in the world discovered that we were threatened by an outer – a power from outer space, from another planet. Wouldn't we all of a sudden find that we didn't have any differences between us at all, we were all human beings, citizens of the world, and wouldn't we come together to fight that particular threat?'

– PRESIDENT RONALD REAGAN, 1988

'A small group of desperate men, who to gratify insatiable ambitions had allied themselves with the thing in the sky, men who were guilty of treason against the entire human species.'

– FRED HOYLE, *THE BLACK CLOUD*, 1957

'Once we have the message … the rest is easy.'

– FRANK DRAKE, 1974

What would be an alien's motivation behind sending a message? Would it be a peaceful call for companions wherever they may be, or a message directed specifically to us? Perhaps the message would be a boast or a kind of monument, possibly the last cry of a doomed civilisation. If we were facing certain extinction, would we send a time capsule of our culture to the stars? Perhaps an alien message would be the prelude to an invasion, or an advanced computer virus designed to be detected by an unwitting species. Once received, it might carry out its real task of infection and invasion. I do hope that those searching for alien signals isolate their computers from the internet. Long shot, I know, but just in case.

We must consider that one way evolution could play out elsewhere is that aliens might not have become wise princes of peace, billion-year-old beings who have left war and aggression behind aeons ago. Is not the lesson of evolution on Earth that predation and fear are fundamental to complex life forms? It does not take intellect to creep up on a blade of grass, but it does if the aim is to hunt and kill other organisms for survival. Intelligence is evolutionary beneficial to predators and it must have happened in the evolution of alien life. The question is, where does it stop, or does it stop at all? Could an alien culture be aggressive and predatory, brutal and cruel with no conception that any other existence is possible? A billion years of stomping out other intelligences, exulting in exclusivity, proclaiming the cosmos as theirs and theirs alone, a hyper-predator with advanced technology in the next star system along from ours, looking at us with what we would crudely call contempt.

The idea that aliens would be peaceful crops up again and again because that's what we hope we would be. Professor Steven Pinker of Harvard has researched the place of violence in human societies, detailing it in his book, *The Better Angels of Our Nature*, in which he concludes that humanity is becoming less violent – an interesting conclusion given the way world events are going. Pinker says mass murder is on the decline compared to the days of Genghis Khan and Adolf Hitler. Will it continue that way for ourselves and for the aliens?

Is all this just wishful thinking, taking the best parts of our humanity into the future? Can we base alien behaviour on human traits? Aliens are aliens. If they wanted to contact us or come to Earth, they must be doing so for a reason. Our experience suggests that expansionist civilisations tend to be more violent because they seek to seize control of occupied territories. Some contend that the galaxy is a very big place with plenty of space for civilisations to occupy without displacing others. Again, these are human interpretations and assumptions. Aliens might not need our solar system, but it might be their habit to possess it and they might just not like us.

If we did detect a signal that we could extract some information from, we must ask whether we can trust it. Let's look at this from another direction. In our tentative transmissions to the stars, how have we portrayed our own species? The Pioneer plaques and Voyager message discs placed on the spacecraft portrayed our good side. There were no images of poverty or pollution, suffering or even nuclear weapons. We cannot honestly declare that we are a peaceful species, or even a just one, despite the aspirations of many. If it came to composing a new message from Earth, there are bound to be arguments about what it should include, and many would want to censor out anything that is negative. Regarding the Pioneer and Voyager messages, one would have thought that a basic element would have been a picture of a naked man and woman – it doesn't get more basic than that. But you will not find such an image in the Voyager message. Politicians vetoed it, one saying that they didn't want to send smut into space. I think that's rather ironic. Aliens that are experienced at first contact might be wise to such tactics, noting with suspicion any culture that omitted what must be ubiquitously embarrassing and horrifying aspects of their history. If we censor our messages to space, then we can surely expect aliens to do the same for their less attractive aspects.

How would we react if we received the alien equivalent of the Memory of the World Register? It is a compendium of documents recommended to be curated by UNESCO for their 'world significance and outstanding universal value'. There is much in it that we can be proud of: ancient languages, music, religious and secular manuscripts, the great works of science and literature, movies and documentaries. There is the *Jikji Simche Yojeol*, an 'anthology of great Buddhist priests' Zen teachings', the earliest known book printed with movable metal type in 1377. But it is not a flattering representation of our species. One of its aims is the preservation of all knowledge and events so that we do not forget what humanity is capable of which can be erased either deliberately or due to the accidents of history – for example, remember the burning of the library of Alexandria.

Alongside the 1789 Declaration of the Rights of Man and of the Citizen, drawn up during the first French Revolution, are testimonies about the Holocaust. Newspapers chronicle the great events both memorable and terrible. The register is the recorded history of humanity. Would we be proud of it? Would we dare send this to the stars? We could explain it all, but would we want to? This civilisation that won't send 'smut' into space? If aliens ever detected our signals what would impress them the most, our honest history or a barrage of teenage Twitter messages? Some talk of tapping into the *Encyclopaedia Galactica*, but I wonder what joys and darkness it might contain. As a civilisation becomes more powerful and spreads itself into the galaxy, would the potential for war crimes increase along with the potential for increased fulfilment?

In November 1985, President Ronald Reagan met Soviet leader Mikhail Gorbachev in Geneva for the first time. It had been seven years since the leaders of the world's superpowers had met and Reagan was nervous. Both sides ostensibly wanted to reduce the number of nuclear weapons. A test of Reagan's controversial Strategic Defense Initiative for protection from nuclear missiles (aka the Star Wars programme) was also forthcoming, which only increased the tension between the two. Their initial meeting was in a villa on the shore of Lake Geneva and was scheduled to last ten minutes but lasted longer. Some officials were worried this was a bad sign, but Secretary of State George Shultz said not to disturb them. The pair emerged and, against the timetable, decided to walk down to a log cabin next to the shoreline. It had comfy chairs and a log fire. Reagan and Gorbachev were alone except for translators. To help break the ice, Reagan asked if the US was being attacked by aliens would the USSR come to its aid. Gorbachev smiled and said: 'No doubt about it.'

The combined forces of the US and the former USSR, formidable as they are, might not be enough. We would not be equals, there would almost certainly be a large gulf between us in terms of time and technology at the very least. In any confrontation one will always utterly dominate the other. However, Sagan often claimed, as so many others have, that aliens who have not destroyed themselves would treat others well and that aliens with warlike tendencies would never survive to contact others or travel between the stars. This might be true, sometimes. Humanity might be considered the alien equivalent of savages, or infidels, with no rights. Thucydides said of the Peloponnesian War that what is right is only a concept between equals and that in reality the powerful do what they want and the weak suffer what they must. We cannot, must not, assume that alien intelligence is benign, and that contact would be beneficial for us. During the Renaissance period of exploration, European conquerors behaved ruthlessly toward conquered peoples. Their ways may not be our ways, their choices may not be our choices. No human knows anything about the nature of alien intelligence.

In my view, Arthur C. Clarke got it so wrong when he maintained that war from space would only be possible between civilisations of comparable development. 'If ships from Earth ever set out to conquer other worlds,' he said, 'they may find themselves, at the end of their journeys, in the position of painted war canoes drawing slowly into New York Harbor.' War is so much more efficient between unequals. Resistance would undoubtedly be futile.

I have a sinking feeling when I read about what Frank Drake considered we could learn from what he termed the 'ultimate social systems' evolved by aliens, a feeling that is deepened when some suggest that aliens might have lessons for us in the development of a benign world government. We imagined many kinds of utopia in the 20th century and the implementation of some has left us scarred and should provide us with a healthy scepticism regarding advice from an alien culture. Have we not learned from our recent experience and the high price we paid for the anti-human attitudes of fascism and communism? Perhaps the aliens will tell us that there are no absolutes, no utopia that can be obtained and that survival is difficult. Beware the words of Arthur C. Clarke when he said that he believed that the exploration of space was a way to 'improve' the human race. In his magnificent novel *Childhood's End*, the Overlords take control of the Earth with a benevolent dictator ruling for our 'benefit'. Only later do their true intensions become clear. They are helping humans to become a better species, to take the next step in evolution, liberating our children and leaving misery, despair and death behind. The pied pipers of space.

Would aliens polarise us? Would some believe that an alien way of organising society has its merits? Would it happen slowly, almost by stealth? In Carl Sagan's book *Contact*, and in the subsequent film, the enigmatic aliens talk of 'small steps'. They have initiated many into the galactic community and know just how much we can take and how often. Like Captain Picard of the USS Enterprise sticking to the Federation's rules of first contact, it is only appropriate when it thinks they are ready. As we consider what an alien contact might bring, being under the control of others would be unacceptable to some. If the aliens presented a better way of life, how would it be enforced? What would the resistance be like? There are many science fiction scenarios about such a situation and the rebel alliances that might ensue. The dream of salvation from aliens, the enlightenment that many who believe UFOs are aliens with a message expect, could be a nightmare.

Such worries are moot, optimists like Sagan have argued. Are we not free to ignore the contents of an alien message if we do not like what it has to say? That would only work if the contact was kept secret from wider society. It would mean putting its power in the hands of the few who would withhold it, manipulate it for the greater good of humanity. At least that is what they would tell themselves. But we could never unlearn it, the genie could not be put back in the bottle. It would be a watershed moment for us. No going back. So, is contact with aliens too much of a risk? Disney or Spielberg-like philosophies are not enough to help us prepare for contact with aliens. Should we stop it from happening as much as we can?

It's possible that in the immensity of space avarice and greed are meaningless. If you are a civilisation that has conquered problems of energy availability and even ageing and biological decay and can have everything you want, perhaps possessions are irrelevant when everything is plentiful. Perhaps what would be important would be the unique, like the tales from a blue planet in a distant star system. But what of the other things? Power might be important. There are computer models that purport to cast light on the implications of cooperation within and between societies. They tend to show that cooperation is good and belligerence is bad. All sides benefit from cooperation says this particular aspect of game theory and it implies this is the way societies, our own and by implication alien ones, will be long-lasting and democratic. If not, then they will collapse, according to Sagan and Drake's views.

This is a conceit arising from the belief about desirable human progress since the Enlightenment. It is also a view that is not universally accepted on Earth, let alone across swathes of the galaxy. Physicist Freeman Dyson said: 'Our business as scientists is to search the universe and find out what is there. What is there may conform to our moral sense or it may not … We must be prepared for either possibility and conduct our searches accordingly.'

Dyson speculated about what he called a technological cancer spreading across our galaxy in just a few million years. The so-called Great Silence – why we have yet to find any transmissions from aliens – might be because we live in a Darwinian universe in which the most aggressive and ruthless prosper, he warned. And they would probably not be biological creatures but machines. How many science fiction stories deal with the development of intelligent machines that outstrip or even turn on their biological creators for their own judgement day?

Somewhere in one of those clusters we talked about at the beginning of this book is a galaxy, probably countless galaxies, where this playbook has come to pass, where the relentless and unstoppable expansion by self-interested machines has laid waste to countless biological civilisations, seeking them out through their radio leakage, deliberate transmissions and technosignatures. Imagine central stellar bulges and spiral arms ravaged by machine intelligences, populating, replicating and patrolling the outskirts of gas clouds and all the time listening. Coleridge wrote that 'and 'mid this tumult Kubla heard from far ancestral voices prophesying war!'. Another explanation for the absence of aliens is the so-called zoo hypothesis, which maintains that we have been put into quarantine, off-limits until we develop further or until aliens deem our isolation should be lifted. There is no way to logically refute this theory. What is the reason for the Great Silence in our own galaxy?

If alien expansion throughout the galaxy did occur and colonies were established, it's unlikely that there would be only one origin of life. What would happen when these beings confronted each other? Would there be conflict between legions of robots, an interstellar war between empires? Sagan said that two expanding alien empires might ignore each other. Why would advanced aliens need planets anyway? They surely would be more than able to live how they wanted in the darkness of interstellar space. Assumptions, assumptions, assumptions. What many SETI pioneers did not realise is that we cannot assume that we are safe because of the great distances between the stars.

Some have speculated that detecting a message will be the equivalent of an invitation to join a galactic club – a network of communicating and cooperating advanced civilisations sharing their knowledge and experiences. Just imagine the wonders of alien races and their views of the cosmos. How many different forms of art and culture could there be? For me, given the distances and the timescales and the limit of lightspeed anything other than a loose galactic club is unlikely. It's a nice idea but it smacks more of what we want our human society to be like writ large in the stars. Perhaps somewhere in the universe civilisations are fortuitously close, or do empires vie for dominance? How would such a galactic club work when its members would be so unequal? How do the junior partners relate to the established ones? Would there be levels of probation, cooperation, coercion, secrecy? Qualifications or initiations? Promotions or executions? And what of the politics of such an organisation. Would there be such a thing as negotiation?

The history of empires is the history of human civilisation; they have been a way of life for most of the peoples of the world for most of the time, as either conqueror or conquered. The general rule is that powerful societies expand. Wherever we have civilisation, we have imperialism. Empires do not want to participate in the international system; they want to be the international system. Would the fear of chaos among the starfields of the galaxy hold things together?

To my mind, it is more likely that popular science fiction stories of interstellar administration concerning empires and intrigues, such as Asimov's *Foundation* trilogy run by humans, *Star Wars* or *Star Trek*, are fantasies. Olaf Stapledon painted a bleak picture: 'By far the commonest type of galactic society was that in which many systems of worlds had developed independently, come into conflict, slaughtered one another, produced vast federations and empires, plunged again and again into social chaos, and struggled … haltingly toward galactic utopia.' In the human case, the task of governance did become more difficult as empires extended themselves. In order to rule vast and widely separated domains, imperial governments generally found themselves compelled to be broadly tolerant of a diversity of cultures and sometimes of beliefs, as long as they posed no threat to their authority. All you have to do is kneel.

So, what would happen if we were asked to become a member? To become galactically civilised? Would it be done by military power, cultural influence, economic advantage? Would we willingly join? The Roman and British empires flourished not only by force, but also by persuading that it was beneficial to be part of an empire. Interstellar empires may depend on means far more effective than any we now think feasible. Even a very advanced technological species would face an apparent fundamental limitation on exerting physical influence: the lightspeed limit. Whether civilisations thousands or millions of years more advanced than our own can overcome this barrier is simply unknown. Before dismissing that idea, it is worth remembering that none of the imperial administrations of the 19th century foresaw that a future technology – the aeroplane – would bring all of the Earth's surface within one day's reach.

What would be the reaction to the news of contact with aliens? Might it depend upon the context? After all a spaceship entering our solar system would illicit a different type of response from the detection of a far-away signal. As a journalist I know what a messy affair such a story would be in the resultant scramble, celebration and trauma.

Captain Cook was aware of the suffering that his visits might cause the Tahitians. 'It would have been far better for these poor people never to have known our superiority in the accommodations and arts that make life comfortable,' he wrote, 'than after once knowing it, to be again left and abandoned in their original incapacity of improvement. Indeed, they cannot be restored to that happy mediocrity in which they lived before we discovered them.'

European contact with the inhabitants of the coastal lowlands of New Guinea produced 'cargo cults'. The natives wanted more *kago*, that is Western goods, but they did not know how they were manufactured so they turned to their ancestors who had bequeathed them the technologies of stone, wood and fibre that enabled them to survive. They believed these ancestors must have also designed the Western technology, but the Europeans had somehow intercepted and stolen what should have been rightfully theirs. To get what they deserved, they stopped working and planting crops, destroyed their own tools and performed rituals to communicate with their ancestors asking them for a new age of *kago*. But the story does not end there. The highland interior of New Guinea was not reached by the initial European settlers. When they did get there in the 1930s, the outcome was much better and the stone-age natives adapted well to their visitors. Within a few decades, they had gained their independence from Australia and developed a thriving economy based on coffee and other crops. Is there a lesson here for us?

People define themselves as much as by what they are as by what they are not. Nationalism on Earth often has an 'us' and a 'them' narrative, demonising the other and sometimes referring to them as subhuman. Rather than seeing ourselves and the aliens as both representations of life in the universe deserving respect, would some regard them as enemies, the ultimate outsider and other, a non-human foe that reinforces our own sense of identity and purpose?

Political leaders could fan these flames no matter what side they might support. Defining the enemy has been a common way of consolidating power, as has aligning with a growing power. The Roman Emperor Constantine used the growing popularity of Christianity to strengthen his base, and much later when the Catholic Church turned on the Protestants, it suited some to support Martin Luther.

Like Carl Sagan's *Contact* and Fred Hoyle's *A for Andromeda*, *The Cassiopeia Affair* by Chloe Zerwick is a message-from-outer-space science fiction novel, but its main focus is the political fallout along with the associated greed, deceit and hatred that would accompany receiving such a message from aliens. This is how the President of the United States announced the existence of extraterrestrials:

> Think of the wonderful things we can learn from our brothers of Cassiopeia. The chemistry of their life processes, the properties of their planet and their planetary system, the Galaxy as viewed from their location – all these possibilities bring tremendous excitement to our scientists.
>
> We believe that Cassiopeian civilisation is technologically more advanced than our own. Their knowledge may enable us to leapfrog over centuries of development. Cassiopeia 3579 might well provide us a sort of cosmic technology program for the emerging Earth.
>
> Above all, our Cassiopeian brothers may have found ways to solve political problems comparable to those which now threaten us on Earth. They may have found ways of managing such problems without participating in a planet-wide holocaust as we have feared on Earth for so long.

What would be the implications of contact with an alien race with superior knowledge? Would the information they have help us solve our problems? Would they help us develop new forms of energy and transport? Would they revolutionise our understanding of our biology, enabling the elimination of disease and even death? When they learn about our progress in understanding the physical universe, would they tell us we have been going down a blind alley and put us right by revealing at a stroke the fundamental theory at the root of all reality? Mathematical conjectures, the elusive Riemann hypothesis, the possibility of time travel – no problem, it's like this…

But would this take away something vital from us, removing our aspirations, perhaps even our dreams? What would it be like if we knew all this stuff and solved all those problems with information we did not strive to obtain ourselves? Would it take away humanity's spark, our drive, as we became self-satisfyingly knowledgeable? Ninety years ago, during the Mars furore, *The New York Times* said it would be better to find out things in our own slow, blundering way rather than to have knowledge for which we are unprepared thrust upon us by superior intelligences.

Surprisingly, we have a small example of what could happen in such a situation because the development of Western civilisation, indeed it's very Renaissance, came about as a result of information obtained from elsewhere, influential information that it did not originate itself. In a way, we have made contact before.

Around AD 500, the Roman Empire collapsed in Western Europe. Societies turned inward, cut off, trade diminished, libraries burned. Scholarship moved into the monasteries. We lost so much, including the library of Alexandria, which once contained Aristarchus' treatise on heliocentric theory, which could have changed the course of astronomy had it survived.

The Hejira of Mohammed began a series of conquests rivalling those of Alexander almost 1,000 years earlier as Islam spread across the Middle East and Africa. Behind the often-violent stormfront of the mujahideen and their imposed religion, Muslim scholars were eagerly seeking manuscripts of ancient works, copying them and sending them back home, and in so doing preserving them for posterity.

Job of Edessa, also known as Ayyub al-Ruhawi, was a Christian and Syrian physician who left us an account of natural science as taught during his lifetime in the early ninth century. He lived in the golden age of learning in Baghdad, which had become a centre of knowledge and commerce. Among its wide and well-laid-out precincts were fabulous libraries containing works from both East and West as befitting Baghdad's central global position. Original works from Greece and India were held in locked chests guarded by suspicious librarians. Translations from Greek and Indian into Syriac, Arabic and Persian started to fill the shelves. For a while there was no greater concentration of books on the planet, and no more exciting place to be, for Baghdad was the largest city in the world and the first to really compete with the glory of Rome at its peak. When Baghdad lost the title of world's largest city, it did so to another city with a magnificent library, Córdoba in Muslim Spain.

It has been said that Córdoba became the abode of the learned, where scholars came to cultivate poetry, to study the sciences and be instructed in divinity or the law. It was 'a necklace strung with inestimable pearls collected in the Ocean of language by her orators and poets; her robes are made of the banners of science,' according to the 17th-century writer Ahmed ibn Mohammed al-Makkari. Hrotsvitha of Gandersheim wrote that Córdoba was overflowing in the seven streams of knowledge.

Toledo, with its own library of ancient works, fell from the Moors to Alphonso VI in 1085, and Córdoba to the King of Castile in 1236, and their vast libraries of Arabic translations of Greek classics was made available to the West for the first time. There were more books in those two libraries than in the whole of England and France at the time. Knowing this, the Italian scholar Gerard of Cremona went to Toledo 'searching for Ptolemy', the Ancient Greek mathematician, and found him. Soon, the works of Galen, Alhazen, Avicenna, Euclid, Archimedes, Hippocrates and others were translated into Latin and such knowledge transformed the medieval outlook. At the beginning of the 12th century, Greek science existed in Arabic, by the century's end, it had been translated into Latin.

Such a storehouse of knowledge held safe during the dark ages changed the development of Western civilisation and gave us the world we see today. Of course, history could have been different and this ancient knowledge might not have been lost and the dark ages not taken place, but this is what happened. The West was given a time capsule from an ancient race.

The Aztec Empire had expanded and flourished and basked in its greatness. Yet that was to change by a new form of conquest. The 500-year empire didn't fall by arms and strategy alone but by Hernán Cortés and his conquistadors, who were said to be gods from over the water. To Moctezuma, they looked like men and he had conquered men before. But this time there was something different, like the Neanderthal encounter tens of thousands of years before, there was a strangeness and some of them were mounted on curious beasts. As these conquerors had slaughtered loyal cities and made alliances with his enemies, diplomacy was the best option, at least at first. He met them near the Great Temple of Tenochtitlán, the symbolic heart of the empire and invited Cortés to stay in the royal palace, but when Cortés was briefly away his deputy took Moctezuma hostage. When Cortés returned, the people rebelled against the Spanish and Moctezuma, who was probably killed by his own side. The Spanish were chased out of Tenochtitlán, but when they returned a year later with reinforcements prepared to take the city by violence, they found it had been ravaged by smallpox. Soon, Mexico's population fell to just 10 per cent of what it was. What happened to the Aztecs is often cited as the fate of an encounter with a more advanced society, but it is not as simple as that. The real destroyer was unknown to the Aztecs and Spaniards alike. Thousands of years ago, modern humans invaded with new ideas, and the conquistadors brought disease.

In 1960, the Brookings Institution studied first contact for NASA and made the following assessment: 'Anthropological files contain many examples of societies, sure of their place in the universe, which have disintegrated when they have had to associate with previously unfamiliar societies espousing different ideas and different life ways; others that survived such an experience usually did so by paying the price of changes in values and attitudes and behavior.'

On such contact, Carl Sagan wrote in 1973: 'The cultural shock from the content of the message is likely, in the short run, to be small.' Later he changed his mind, saying 'there is a significant potential for culture shock'. Undoubtedly, many would be shocked by the news of the existence of aliens. We have already considered the Neanderthals and the Aztecs. These were two advanced societies, each master in its own domain, but both were to change irrevocably the moment they met outsiders, and both knew that only one of them would survive. Is alien contact to be feared or welcomed? In 1976, Frank Drake said that we need not be afraid of interstellar contact, for unlike the primitive civilisations on Earth which came into contact with more advanced technological societies, we would not be forced to obey – we would only receive information. Is this a naïve view? Is information dangerous on its own? I think it is.

Some welcome what others view as a catastrophe. 'I would like to see our culture merge with the other one,' said German astronomer Sebastian von Hoerner. 'If there is a galactic culture, a galactic club, then why not join it? We would lose our own present culture, or what we think is a culture, and merge into a larger way of life. This is the only way it should go.' Astronomer Paul Horowitz thought that contact would be 'the greatest event in the history of mankind'. Sagan shared that view: 'The scientific, logical, cultural, and ethical knowledge to be gained by tuning into galactic transmissions may be, in the long run, the most profound single event in the history of our civilisation.' Sagan added: 'It is certainly possible that the future of human civilization depends on the receipt and decoding of interstellar messages.' Arthur C. Clarke said it might be the most devastating event in our history. Stephen Jay Gould said it would be cataclysmic.

These are dangerous ideas. Some maintain that contact can regenerate societies, look at what happened to China in the 19th century, they say. Perhaps we would become depressed at the state of our technological backwardness. Frank Drake was more optimistic. 'Some eminent people say it will be terribly depressing, that we'll feel ignorant, and they predict a planet-wide inferiority complex. My take is that it would have the opposite effect. It could motivate us to think that if we worked hard we could be as good as them, motivate us to make progress much more quickly than we are ... we all have been exposed to minds and accomplishments greater than our own,' he argued. 'The result is more often inspiration rather than depression.' But how can he be sure? Sagan and Drake thought that contact with aliens 'would inevitably enrich mankind beyond imagination'. 'Searching for other life in the universe is not an unnecessary luxury,' Drake maintained, 'but an essential component of forging a better life for Humankind.' Would we gamble the wellbeing of humanity on such an assumption? It is a narrow debate usually concentrating on the scenario of a slow exchange of radio messages across hundreds of light years with no physical contact which tends to be the most optimistic scenario. There is time for the information to be studied and released. The distance mitigates any threat. It's just information. It's the Ancient Greeks all over again.

In Carl Sagan's *Contact*, religious fundamentalists oppose communications with aliens, so much so that a suicide bomber destroys the alien-designed portal. Will some see aliens as the devil from deep space and the punishment for our sins? Some believe our civilisation will be shaken to the core by the news. Carl Jung said that finding a far-more advanced civilisation would leave us 'without dreams'. Others have said that it will cause a fuss for a while and then, given the vast distance of space between us, it will fade and we will get on with things as normal. That may be the case for most of us, but there are some aspects of society that will never be the same.

When we find aliens, and if we are able to exchange information, I would, at first, not be so interested in their scientific discoveries, as much as in what makes them different to us. How did they evolve? What do they look like? How do they organise their society? Do they have morals and a sense of right and wrong? Most of all, do they consider themselves to have souls? Do they have a God? Some religions have seriously considered the implications of finding intelligence in space, others have worked hard to avoid the issue altogether. The fact is it raises troubling questions that strike at the very foundations of some faiths.

The mathematician and theologian Bishop Ernest Barnes said that finding other intelligent beings more advanced than we are would challenge our self-image as the chosen people of an anthropocentric God. 'If God only realises Himself within an evolutionary progress,' declared Bishop Barnes, 'then elsewhere He has reached a splendour and fullness of existence to which Earth's evolutionary advance can add nothing' Would the 'reins be torn from our hands', as psychologist Carl Jung thought when 'we find ourselves no more an intellectual match for superior beings than our pets are for us, to find all our aspirations outmoded, might leave us completely demoralised'?

Islam may welcome them. The Koran actually mentions life from outer space, and some Muslims have said it is the height of conceit to suppose that God created the vastness of the universe just for us to enjoy when we will never see the vast majority of it. Buddhists and Hindus with their philosophy of oneness and unity will have no problem accepting aliens as part of a cosmic brotherhood of consciousness.

It is Christianity that will find it hardest to adapt. There is a particular hard line of Christian thought that would deny aliens, however advanced, wise and serene they may appear to be, the same status as us. There is God and the angels, who are spiritual beings without form, there are humans, who have a soul, and there are the animals. Aliens, because they are not human, are animals and therefore do not have the same moral status we have.

Some scholars think aliens have souls. The Vatican astronomer Father Guy Consolmagno believes they do: 'Any entity – no matter how many tentacles it has – has a soul.' He also adds that he, for one, would baptise an alien into the Catholic faith if they asked, implying that once baptised they would be spiritually the same as humans as all souls are equal before God. Indeed, some say that if we found aliens, it would be the church's duty to convert them, in the same way the church travelled to the New World to save heathens.

The problem is that mankind and the Earth are privileged in Christian theology. Christians have a personal God and a personal relationship with Jesus. They believe that out of the immensity of space and time, God, the creator of everything, has singled out humans to be saved and has sent his son to redeem us. The question is, therefore, has God done the same for the aliens? Opinion is divided.

Father José Gabriel Funes, the former director of the Vatican Observatory, has stated that in a cosmos with possibly billions of intelligent life forms, only humans have been saved and that God's son only came here in a unique and unrepeatable event. Jesus did not suffer on other worlds. However knowledgeable the aliens may be and however seemingly good and wise they have become, they were not visited by God's son, who did not die to redeem them. According to Father Funes, for the aliens to have fellowship with God there will have to be another way. The Catholic commentator Joseph Breig wrote in 1960 that there can be only one incarnation, one mother of God and only one race into which God has poured his image and likeness.

Some have suggested that God's son could have visited other civilisations in space and redeemed them. In 1913, the poet Alice Meynell wondered in what guise Christ trod the Pleiades, asking if there were a million gospels and a million forms of Jesus among the stars. Others strenuously point out that Christianity is Christocentric, not anthropocentric. The theologically difficult point is not the existence of aliens but the right place of such beings in the history of salvation. Christianity will have to adapt or cling to its chosen status and consider aliens as infidels.

One day we may converse with an alien. A future Pope and an alien may discuss questions of faith. But before that takes place, the search itself forces us to confront difficult questions about our beliefs. If we are made 'in the image and likeness of God', are creatures from the depths of space and who bear us no biological relationship also made in his image? Arthur C. Clarke once wrote that a faith which cannot survive collision with the truth is not worth many regrets. Aliens might share some religious concepts with us, such as the existence of a supreme being or the deliberate creation of the universe. Perhaps our concepts of God are a subset of possible views. Any alien religion could not agree with all of the diverse religions of Earth, so someone will be annoyed.

The Christian church has lived with changing times from the Ptolemaic view to the Copernican view. It might be that resistance to change came not from the theologians, but from society as a whole. Great religious leaders take for granted the worldview of their time, usually on a very popular level. Christianity survived Copernicanism, Darwinian theory and Marxism. The adjustment will be less wrenching for Eastern religions that teach salvation through individual enlightenment. 'If science proves some belief of Buddhism wrong,' said the Dalai Lama, 'then Buddhism will have to change.' Religion has had to come to terms with evolution, seeing it as God's work. Is a melding of science and theology feasible? Many religious leaders acknowledge the validity of the scientific approach, although perhaps not for all questions. If our religions and those of an alien are incompatible, do we have choices other than adopting their beliefs or rejecting them completely? The common ground may not lie in the intellectual heights of theology, but in the practical world of ethics, the way intelligent beings in one society treat their counterparts in another.

Would we even know? Would not the very existence of a signal from an alien be kept a secret? If there is lots of information received, Frank Drake urges caution: 'You'd better take a close look at the information to see if it would appear threatening to anyone, and make a judgement as to just what you say.' So, let's be realistic. Such a signal might not be made public, but then others would want to get in on the act. The information would be valuable. Would it become the most highly classified information in the world?

There have been, and undoubtedly are, many secrets in science. One was over the test detonation of the world's first thermonuclear bomb in 1952. A concern at the time was that the explosion could ignite a runaway chain reaction that could incinerate our atmosphere and all life on Earth. Obviously, that did not happen, but top-secret talks had taken place beforehand to assess the risk, and the question of whether the experimental explosion should take place was never placed in the wider world's hands. Another example was some people's fear that the Large Hadron Collider (LHC) – the powerful particle accelerator at CERN on the French–Swiss border, which discovered the Higgs boson – might accidentally create a black hole that could swallow the Earth from the inside out. That debate was held in the public domain, ending in the courts when judges ruled, having seen the scientific evidence, that the LHC could be turned on with negligible risk to the Earth.

There are precedents. In 1967, the US Department of Defense's Vela 4 nuclear test monitoring satellite recorded a flash of gamma rays that did not come from a nuclear test. Instead, they came from the opposite direction, from deep space. The data was analysed in 1969 but not made public until 1973. Astronomers now know them as gamma-ray bursters. It's not just the military but academics as well. In 1967, astronomers at Cambridge University detected pulsing radio signals. One of the first thoughts about their origin was that they might be artificial. But instead of calling a press conference, they withheld their results for months. They later said their caution was justified as the signals came from a new type of natural object.

In 1995 to 1996, researchers looking at a sample of Martian rock concealed their findings that they might have found traces of past life. A meteorite picked up in 1984 from the Far Western Icefield of the Allan Hills region of Antarctica was later identified to be from Mars, one of a few that have made their way here after having been blasted into space by an impact, in its case some 17 million years ago. It was taken to the NASA Johnson Space Center in Texas for analysis. A small number of scientists noticed some things that were strange about this rock. When viewed under a scanning electron microscope, small structures were seen that were hypothesised as being nanobacteria. If confirmed it would have been the first solid evidence of extra-terrestrial life. For over a year the researchers kept the discovery among themselves, shuffling paperwork and assignments, to keep their work away from attention. Eventually, they arranged a meeting with the head of NASA, Dan Goldin. It was scheduled to last less than an hour but they stayed with him all afternoon. When it was over, Goldin was beaming and apparently hugged the scientists. After they left he asked his assistant to put in a call to the vice-president.

The research was due to be published in the journal *Science* in August 1996 but leaked a few days before. Looking back, the research is intriguing and still inconclusive, but the case for alien life hasn't been conclusively made, there are other explanations and we need more rocks from Mars to examine which should get here later this decade via an international sample return mission.

Any rumour of an alien detection would spread rapidly across the internet no matter how circumspectly the astronomers who found it behaved. They would have to have the signal confirmed by others using a different radio telescope, an act that would be known by many: the astronomers themselves, along with their students, but also the observatories' technical and administrational staff as well. No matter what protocols were put into place, finding an alien signal and its verification would rapidly become public news. One astronomer who was involved in searching for alien signals using the Arecibo radio telescope in the 1990s told a story about the finding of a potentially interesting signal that was highlighted for follow-up investigations. It turned out to be interference but before that was determined the observatories control room received a phone call about the suspect signal from a *New York Times* reporter. In the main, scientists are averse to secrecy about such a topic. One could cynically observe that the first scientist to make an announcement would become world-famous and all that would result from it. The competition and the pressure would be intense. This type of scientist is about the worst, or best, person when it comes to keeping such a thing under wraps. As a former science reporter, I can tell you that it was not unusual to receive a phone call from a scientist who was about to embark on a round of SETI observations just to let me know their direct line in case of any developments.

Given the pace of communications in the modern world, a credible detection confirmed by an independent observatory would splash around the globe in moments followed by a deluge of media demand for quotes and statements. Many news outlets would get the details wrong but the gist of the story – the possible detection of intelligent life in space – would be a sensation and lead news bulletins everywhere. There would be a hastily arranged press conference giving sparse details and profound quotes. Pressure would mount for reactions from heads of governments, the Pope and the leaders of other religions.

As private sources fund more of the searches these days, what happens if those who invest millions of dollars want to embargo the details of the message? In an adjacent situation, the full contents of the Dead Sea Scrolls were not published until 54 years after their discovery. One can imagine scientific priesthoods that decide what the rest of us should know about messages from the skies. We cannot assume that the search for extraterrestrials is immune from the ancient motivations of egoism, power and greed. Decisions that could affect the welfare of the human species might be made by small, non-representative elites.

In 2013, the US Supreme Court struck down patents on two human genes associated with breast and ovarian cancers – BRCA1 and BRCA2. Justice Clarence Thomas said that DNA 'is a product of nature and not patent eligible'. The decision invalidated patents held by Myriad Genetics, allowing other companies to develop kits for testing that determine what your BRCA genes say about your risk of getting cancer.

The principle being debated in this case was the idea that natural phenomena as well as observations of laws of nature and abstract ideas are unpatentable. If they were patentable, then many were worried that the world's genetic resources would be owned by vast corporations with scientists feeling that they were always under the cloud of legal action. After the ruling, lawyers were instructed to fight the case and seek to overturn the Supreme Court judgment. What is the incentive for research, they argued, if it's not possible to obtain any commercial benefit from its products? Before the ruling, more than 4,300 human genes had been patented.

The interpretations come from a single provision in US law, which stated: 'No implicit or other judicially created exceptions to subject matter eligibility, including "abstract ideas", "laws of nature", or "natural phenomena", shall be used to determine patent eligibility under section 101, and all cases establishing or interpreting those exceptions to eligibility are hereby abrogated.' The key word here is abrogation, which means the formal nullification of a law through an act of Congress – as in, Congress overruling the outcomes of the Supreme Court. The ruling basically meant that whatever legal approach the courts used in the past wasn't going to be used any more.

It all goes to show how hard it can be to figure out what is an abstract idea? What is a law of nature? And what is meant by 'natural phenomena'? It also, lawyers tell me, updates the legal definition of the word 'useful'. To be eligible for a patent, a discovery or invention has to have an identifiable benefit made possible only through human intervention.

So, the question is, who owns the alien message and the information within it? If it is a company or university, they might want to patent this intellectual property. Billionaire entrepreneurs might compete to get access to alien ideas and to monopolise its commercial value. The lawyers at least will ask why millions of dollars were invested in SETI if you don't market the information that it discovers. John Lilly wrote in 1961: 'If and when interspecies contact is made … it may be that we shall encounter ideas, philosophies, ways and means not previously conceived by the minds of men. If this is the case, the present program of research will quickly pass from the domain of scientists to that of powerful men and institutions.'

A lot has been written about how we might compose a message to aliens and most, if not all of it, is essentially useless. The prime function of a message is to be noticed, the rest is up to interpretation no matter how much debate goes into how 'simple' it is and any fundamental universal qualities it might have.

Some SETI researchers say we will have lots of time to compose a reply should we receive a message from aliens. Not everyone will agree. Some may want to reply urgently and there would probably be a barrage of signals sent into space, and most, if not all, of which would not reach their destination, probably thankfully. In a way, having humanity speak with many voices may reflect the current state of our civilisation, but what would the aliens make of it? Thousands of uncoordinated messages from Earth. Imagine what they would be like.

There might be a reply sent from the Vatican, or from independent Mullas, from the Chinese Communist Party, from the Trump Organization or the Planetary Society. There would be no rational dialogue. 'Listen, brother, this is the only true message from Earth,' might say some messages. Or: 'Enter our competition now for the chance to send your message to the aliens.'

The real issue may be whether superior transmitters, such as large radio telescopes, would be used to send messages. Nearly all of those facilities are funded by governments. Depending on the nature of the contact, policy makers might have an opportunity to make conscious decisions about sending powerful signals from Earth.

How do we express our values when we do not agree among ourselves? Do we need to draft an outgoing message in advance of a detection, as a way of focusing our thinking? Such a message could be reviewed by an international body that would decide whether to send it. This exercise would have implications reaching beyond the immediate issue of message content. Perhaps only when we begin composing a reply message will we begin to understand the range of issues involved in deciding on the content. Building a global consensus on how to represent humankind would have a significance reaching far beyond the immediate issues of contact.

Let us move forward and contemplate the future for life in the cosmos. And we begin our new journey with a visit to a new museum – the Museum of Alien Life.

ARBITRAGE

'There are infinite worlds both like and unlike this world of ours ... We must believe that in all other worlds there are living creatures and plants and other things we see in this world.'

– EPICURUS 341–270 BC

'There's no limit to strangeness ... the most likely form for E.T. is something we never imagined.'

– FREEMAN DYSON

Will life appear anywhere it can? Some believe that life requires such special conditions that places like the Earth must be very rare indeed. This is called, naturally, the Rare Earth hypothesis. It states that our planet is a one-in-a-trillion fluke. I do not believe this theory. I prefer the idea that life could evolve in any habitat capable of supporting the required complexity of organisation. The range of environments for life that the universe offers is huge, and we do not know the limits it places upon living creatures. We have already taken life on Earth as our first reference point and I ask how far can we go? Despite being conditioned by our own biota, it is only our imagination that offers us a guide to what kinds of life might be out there. I will use that imagination and wonder if somewhere out there exists the equivalent of our Museum of Life on Earth. What for us is a single tree of life could be just one growth in a vast forest of possible life. Where there is energy flowing, nature's universal principle of arbitrage unleashes the possibility of life, and there is energy flowing everywhere.

Without exercising our imaginations too much one can speculate about what could be possible among 50 billion water-rich worlds, on stable planets, constant in conditions at an equitable distance from their star, or on planets driven by the orbital rhythm of double and tertiary stars, or on ice moons locked in the tidal grip of gas giants. What creatures would be sculpted by a different pull of gravity, different light levels or living in alien soundscapes? The same laws apply everywhere, using the same atoms but using the hypothecated millions of possible alternatives to DNA could produce new plants, creatures, something like humanoids perhaps, or perhaps not. Anthropologist Loren Eiseley said: 'Nowhere in all space or on a thousand worlds will there be men to share our loneliness. There may be wisdom, there may be power; somewhere across space great instruments ... may stare vainly at our floating cloud wrack, their owners yearning as we yearn. Nevertheless, in the nature of life and in the principles of evolution we have had our answer. Of men elsewhere, and beyond, there will be none forever.'

Where will this life be? Our planet of water and oxygen is only one possibility, ice moons are another. There are also the surfaces of other worlds, the clouds of gas giants, the interiors of stars, the vast interstellar spaces or clouds of molecules, magnetic fields and the surface of neutron stars as speculated by Robert L. Forward in his novel *Dragon's Egg*. His 'Cheela' live on a neutron star in 67-billion-G gravity. Forward provides a very detailed description of the evolution and social structure of the Cheela.

As Eiseley pointed out, the vast majority of aliens will not look like humans. We are descended from lobe-finned fish that have bequeathed us two arms, two legs, five fingers, knees and elbows and teeth which were once scales. Aliens will be different, although bilateral symmetry and jointed limbs are probable in my view. They certainly won't look like humanised reptiles or insects even though some will have their sensory organs on one or more appendages. Their biochemistry will differ from ours but even if it was similar it wouldn't result in many similarities.

So, let us visit a museum of the adjacent possible, an extension of the known phase-space occupied by Earth life. Behold a sensorium of sentience.

The architecture of such a museum of cosmic life would not be designed just for bipedal, laterally symmetric life forms like ourselves. It is multi-functional, able to adapt to the sensory and intellectual requirements of its visitors and includes pathways for chlorine-breathers and those creatures that can modify their senses at will. In many aspects, its organisation would not fit our logic, but let it adapt to our needs and just for a moment walk with me among the wonders that could be collected from a billion stars and times, and many other places as well.

Symbolically, the Museum of Life in Space is not tied to a star or planet circling a star. It has taken its cue from the many life forms that chose interstellar space as their habitat, thus demonstrating it owes no allegiance to any one species, collaboration or lifestyle. It is orbiting a nomad world moving in the vastness between the stars ultimately heading out of the galaxy. Many species have come to appreciate being far away from the perils of a planetary system, with space for anything.

An appropriate first visit for those who do not wish to delve directly into the diversities of intelligent or conscious life, or the debate about what those terms actually mean, would be to the continent of pre-life to witness a variety of autocatalytic chemical processes self-complicating themselves into something that has the potential for life to develop, though different perceptions position the line between life and non-life in very different places. Only a section of the pre-life exhibits require carbon chemistry and there is a popular display of pre-life based on non-chemical substrates, such as vortex-based life involving multiple vortices of swirling magnetic fields with their differing properties representing something like genetic information. Some autocatalytic mixes of chemicals compete with others for energy and resources, demonstrating features that are familiar in many life forms.

Part of the museum shows how life can spread among the stars, explaining why at a molecular level some regions of a galaxy can have biota that share a common origin. The displays have been optimised for your senses and you see flashes of light, each signifying the emergence of life, spreading down the spiral arms of the galaxy. It is an interesting cosmic experiment; take the raw materials and see what happens in different environments. Once life has originated on one planet, or elsewhere, there are ways it can spread far and wide. After asteroidal collisions, microbe-bearing rocks can be flung into space and transpose life between worlds or even planetary systems. The speed required for ejection is such that a rock may be flung from its home planetary system altogether and take only about 100,000 years to reach a nearby star. Some federations have been based on common origins, some destroyed because of them.

On our planet, bacteria aged 1 or 2 million years old have been revived from arctic ice. More controversially, some scientists claim to have revived spores from the guts of bees that had been entombed in amber for over 20 million years and even more remarkable is the claim of re-animated 250-million-year-old bacteria from a salt crystal mine in New Mexico. There is a side display on the topic of panspermia, of which our parochial definition states that life did not originate on Earth but was seeded here from somewhere else. If we find aliens on planets around nearby stars, it would be fascinating to know if we are cross-pollinated. On one display it is suggested that when the galaxy was young, some as yet unidentified species carried out such an experiment to see what evolution such a dispersion would produce.

Another display area you encounter poses the question of whether galaxies may have preferential habitable zones where the conditions favour life. Some species contend that the centre of the galaxy is a harsh place for life to develop because of the devastating radiation from its frequent supernovas as well as the gravitational disruption from the supermassive central black hole and stellar encounters. On the other tentacle there are more stars born in these heartlands than anywhere else. Moving away from the centre towards the galactic rim there is a tendency for lower abundance of the heavy elements required for life.

For our type of life – carbon-based, bipedal, bilaterally symmetric, endothermic – there is much in the Museum of Alien Life that we recognise, for the story of our form of life is a common fugue performed on myriad worlds. There are so many rough copies of life on Earth, some better than others. On some alien worlds something like dinosaur creatures never went extinct, producing an intelligent species. Changes in gravity, in the type of star and the accident of moons, all add to the variety. Such creatures from small rocky worlds, it is recorded, tend to be aggressive.

Moving on from pre-life, we see a dim landscape of complex structures, rather like the stromatolites – fossilised colonies of microorganisms – of early Earth, this is the view of an under-ice ocean, such as Jupiter's moon Europa in our solar system. Colonies of filamentary bacteria are patrolled by so-called grazers exploiting the temperature difference between the cold water and the superheated water from thermal fissures. On some worlds the colonies form vast towers that in turn affects the ocean circulation, creating eddies and whirlpools where nutrients are concentrated, and where creatures lie in wait. The under-ice oceans evolved few species that have developed technology, most stopped at variations of Europa's anglerfish. But there are some examples of intelligent species that developed in oceans, under ice crusts or even beneath perpetually cloudy skies that found a way to leave their home worlds and travel to other worlds, experiencing a perspective that was new to them. In a display of the atmosphere of a gas giant planet you see octopus-like 'blimps' drifting in the hazy upper-cloud levels, looking for thermals that bring detritus up from the grimy lower levels. Every so often, one of these flimsy blimps explodes, scattering its seeds like rain that falls to denser clouds where they dine on aerokrill before their sacs swell and buoy them upwards.

Life needs a code of some sort, a way to store its basic information, on Earth we call this a genome and there is a vast collection of genomes in this museum. Not many of them are coded in DNA molecules likes ours, many don't use molecules at all. There is a sign telling visitors not to confuse a genome with life as life is much more than a string of code in whatever language it is written in, perhaps in the same way that markings on a musical score is not music.

Some alien species deliberately tinkered with their genetic code. You will recall, our DNA is a double helix: two complementary strands, each with a sequence of bases chosen from a set of four. The helix has a central core of sugar molecules, and each triplet of bases encodes one of twenty amino acids, which in turn make proteins which do all the fundamental processes of life. Some species extended the range of their DNA, adding new components that produce new proteins. This has already been done on Earth with new DNA bases being inserted in bacterial genomes. It seems that the usual DNA bases are nothing special and could easily have been replaced by others, as this display shows. Redesigning life from its base molecules up became a compulsion for many species and most survived the outcomes of their experiments. In some regions of the galaxy, it is not possible to determine if certain species arose naturally or if they were the result of some ancient design.

All around are strange textures of life and strange tongues to be found, alien satisfactions and beings whose lives can only be hinted at by their remains, stratifying the borders of other dimensions. Biology is flexible. Scientists on many worlds had suspicions that somehow the universe was just right for life, that if things had been just a little different, the forces of nature, its constants, such as the fine structure constant or the mass of the electron just a shade heavier, then life would have been impossible, certainly our form of life. On your tour of the museum, you stop by a narrator who tells you this wasn't the case. Using one of her composite bodies she tells you that it is life that is fine-tuned not the universe. In the vast majority of universes, life finds a way to exist (this she had on good authority, experiments had been performed). Our universe is nothing special. Even the universes where conditions that were so strange that one would think life was impossible hold their surprises, she adds, flowering and dissembling as you take her leave.

In one section of the museum, it is possible to run an experiment, or perhaps a kind of game, simulating the clash when two alien biomes come into contact. You can choose two low-level biomes and watch as herds of fractally dividing red crystal-like organisms move towards clouds of swirling green jellies diving through them. The outcome is unknown as the roles of predator and prey, parasite and symbiont, have not yet been established in this first contact situation. Then a huge creature comes into view that seems to turn itself inside out to mop up the aftermath of the encounter. It is surrounded by swarms of smaller creatures, all of which seem to be decaying as if being consumed from within. More creatures appear displaying a behaviour unpredictable by just studying a genome. After a while, everything fragments and decays into a sludge. Neither biome seems to have triumphed and the display says it's too early to judge, telling us to come back in a million years.

It is the conscious, intelligent species that fascinate you most. Part of the museum is devoted to those that have achieved great things. In the 1960s, the Russian physicist Nikolai Kardashev speculated about advanced civilisations and classified them into four types depending upon the amount of energy they had at their disposal. A K0 uses very little power, think Neanderthals. We are currently approaching a K1 species using the resources of our planet. The next stage, K2, would be able to use all the power coming from their star perhaps building surrounding structures able to absorb this energy. These are called Dyson spheres, after the physicist Freeman Dyson, although he wasn't the first to come up with the idea, which can be traced back to Olaf Stapledon's 1937 science fiction novel *Star Maker* and a 1929 proposal by J.D. Bernal. Dyson spheres are held in collections positioned in adjacent space. Eventually, the outside of the Dyson sphere would become hot and potentially be visible as an infrared source. It would be highly visible, Dyson once told me. Our telescopes do not have the resolution to see a Dyson sphere. What we can look for is a star with an excess of infrared radiation. The problem is that there are a lot of natural objects that look just like that. Astronomers have searched the database compiled by the Infrared Astronomical Satellite (IRAS), which was launched in 1983 as the first infrared space mission. Although it operated for only a year, IRAS discovered about 350,000 new sources of infrared radiation, including the first ever detection of planetary systems in the making, in the shape of dust discs around the stars Vega and Beta Pictoris. K3 civilisations would be able to use the entire resources of a galaxy. As far as we can tell, we can see no evidence for their existence. These are profound observations. One can imagine that a K2 society might colonise each star system it encounters, building a Dyson sphere and so on until the galaxy is filled with Dyson spheres, but a comprehensive search of 100,000 galaxies found no evidence of this.

If it's possible, then a K3 civilisation could master space and time. Some speculate they could spawn baby universes, and travel through the multiverses, so perhaps it is understandable that we haven't detected them. In a survey of galaxies, not a single one displayed any evidence for a K3 civilisation. Even so, the result is pretty clear: if K3 civilisations exist, then they are rarer than one in 100,000 galaxies.

In the archives of the museum is a story typical of an emerging alien species. They leave their home planet, visiting nearby ones and eventually reach their Kuiper Belt. Here they can hop on to objects whose orbits transition between the Kuiper Belt out to the so-called Scattered Disc, a realm of icy cometary bodies that were pushed outwards when the outer planets changed their distances from their star. Moving from frozen world to frozen world, they eventually reach their own Oort cloud about a light year from their home and make the hop to the Oort cloud surrounding a nearby star. Oumuamua demonstrated that it is possible to span the distances between stars.

Sex is found often in alien species, though some have abandoned the idea as they gained complete control of their environment and biological functions and decided that evolution must be planned and controlled. Creatures from rocky worlds, gaseous planets, molecular clouds, ice floes on comets and even on the surfaces of neutron stars use sex as a means of procreation and as a driver of evolutionary change through the shuffling of genetic information. This leads to the question: are two sexes universal? Probably not as this isn't even universal on Earth.

Some civilisations have as many as twenty sexual variants. Many creatures have what we on Earth would crudely define as male and female sexual roles but expressed at different points in their lifespans, sometimes instinctively, sometimes by choice. In one civilisation shown as an example, there is only one male and if he dies the next largest female becomes a male, sometimes it's the other way around. This is a process that's been observed on Earth. Sometimes there are several sexes and several generations involved, see associated displays on foremothers and hindfathers. Perhaps, like on Earth, sex can come in many guises, some species chose mates that they are related but not too closely to avoid genetic problems. Others will only mate with close relatives. Lots of species do away with sex altogether and self-fertilise. On Earth, the cockroach *Pycnoscelus surinamensis* and the stick insect *Carausius morosus* have no need for fertilisation as females produce females from eggs that don't need sperms. Some alien civilisations are based on an extrapolation of a situation that is found on Earth in the myth of an Amazonian tribe of women who raided local villages for their males. On some worlds genetic material from males is needed to trigger the development of the next generation but this does not get included in the genes of the offspring.

We next reach the labyrinths of gestalts – creatures who have become more than individuals, some of which have an accumulated consciousness. Some, like the Earth's ants and bees, subsume themselves into a collective whole, each have their role and together they enable the collective to act in many ways like a single, flexible organism that stays complete while benefiting from the expendability of the individual. Humans from some societies would understand these creatures. But what of the species that chose such a way of life? There are the brash entities who melded, some using technology, to pursue a single purpose; the pleasure explorers, merged minds dedicated to solving the mathematical hypothesis or exploring wave-particle duality. There are some combined consciousnesses that design higher versions of themselves and then transfer themselves to it. Then there are the subtle architects of galaxies and those who none understand and whose existence is only reflected in our dimensions.

What histories, proclivities and desires would make a civilisation opt for a gestaltic way of life? How would individual consciousness interact closely with others and what would be its emergent properties? In the catacombs of the museum there are creatures that once stood as individuals the way humans do, passing on their cultural information firstly by example, then via language and later social means, such as through education, libraries and the internet. Now they are sharing thoughts and emotions, some appearing novel to human sensibilities, others sinister yet still enticing. For all those gestalts who sought a higher purpose there are those that dedicated themselves to domination. It is no wonder that so many species decided to keep quiet.

Then the museum presents the mysteries. When you visit the isthmus of the disappeared, many of its creatures accost you. They are the speakers for the dead. It is a place of incomplete exhibits shaded with half-stories, tales of those who arose, perhaps flourished and then vanished. Many of these stories are tragedies and involve creatures of great achievement or great promise who were destroyed before they acquired the skills for long-term survival. Fragments taken from the databases of ancient space probes, decaying libraries on abandoned planets, time capsules tucked into folds of space time and beams of entangled particles sprayed into the cosmos are the final testaments of dead races destroyed by an asteroid strike or by a supernova explosion, genetic instabilities or social trauma leaving behind only stories of what might have been, along with old myths and curious figments of strange artwork.

Represented in various places in the museum are those who came first, creatures who acquired sentience before most others. There are speculations about where they went and the artefacts they left behind, perhaps deliberately.

There are low-gravity worlds that set gentle creatures free to roam the skies and grow tall, allowing life in an atmosphere to exhibit many of the three-dimensional traits of life found in the water worlds where buoyancy offsets gravity. The tallest creatures in the museum reach for their sky in a spindly way, encompassing civilisations for whom the horizon is in every direction. Shimmering multicoloured leaves seem to blow past, fluttering and murmuring. Conversely, the creatures that live on high-gravity worlds that require large amounts of energy to exist are thick-set and strong. In general, intelligences living in such places are late in developing spaceflight, if they have developed it at all.

There are many places in the museum that we humans will never understand. The Isles of Strange Emotions is for us an unsettling place. For some, the museum will shatter illusions about the primacy of love and human 'goodness'. Is it fear and survival rather than altruism that is spread much wider across the cosmos? Do compassion and tolerance have an ultimate survival valve? Humans and many other civilisations will have what we call morality, but it is not always something we would understand in other worlds. In this section there are examples of emotions we know nothing of, or at least not for a million years or a million experiences. There is also an exhibition space where compatible species can upload their emotions.

You could visit an exhibition set beneath the ice of a water world, looking at creatures reminiscent of anglerfish on Earth with luminescent lures dangling in front of them hoping they will attract something, food or a mate, out of the eternal darkness. One of the common themes of the universe is that many of its creatures carry out their lives in darkness, conversing with those they never see, as is the case for many oceanic creatures on our own world.

Societal organisation is explored on one of the museum's adjoining moonlets. There are avenues of totalitarian civilisations and example upon example of warrior species, some gestating in their honeycomb-like lattices. Here are displays devoted to the great dictators of the galaxy, usually from long-lived species that existed before the development of swift interstellar travel.

The curators of the Museum of Alien Life want its visitors not to become obsessed with molecules as the basis of life. There are the civilisations who have retreated, or perhaps they would say advanced, into virtual existences. Some of them believe the universe itself is but a simulation and they are but a dream within a dream. They call themselves the most knowing creatures in the galaxy and they usually leave others puzzled. Others recreate the past, tending their virtual brains.

Life in this museum has a vitality that is unexpressed on a planet like Earth. Some civilisations have become obsessed with ghosts and what might have been and have taken their genetic code – their equivalent of human DNA – and created all possible forms of life with it. For comparison, most possible humans will probably never be born and would outnumber all the stars in many galaxies. There are many strange obsessions to be found in the precincts of this museum. Some societies want to record everything that ever happens to them. There are those that live on planets circling nomad stars – planets set adrift between the stars and the galaxies – torn out of their orbits by stellar and galactic encounters. Some will have thick atmospheres, and some will have under-ice oceans that can sustain life with no need of a star.

Next you visit the segments on black hole life said to be increasing in popularity, especially as long-lived or immortal species look to the future. As stars age and eventually die out altogether, life will begin to rely on black holes. Some life forms live in the vicinity of the event horizon surrounding a certain size of black hole. If the conditions are just right – the black hole has to be spinning – its gravity can focus and amplify the cosmic background radiation so that any orbiting debris has a stream of useful energy occasionally interrupted by shadows. The resulting temperature differences drive thermodynamic processes used by life. Living at the innermost stable orbit of a black hole, they are deep in a gravity well and for those life forms time travels at a different pace. They can look out at the rest of the universe and see everything happening swiftly. For them, time is shortened compared to those living away from a black hole's gravity.

There are some worlds that rely on black holes in a different way. As material is sucked into a black hole it forms a so-called accretion disc, which glows white hot radiating prodigious amounts of energy. Many billions of miles away are worlds and civilisations that rely on its radiation in the way that others rely on a star's radiation.

There are creatures who live in stars. Carbon atoms are most often used as the basis for the molecules for life, although not exclusively as silicon can also form bonds though usually they are not as resilient as carbon ones. In certain circumstances, such as in the dense, high temperature environments of white dwarf stars, so-called firebirds have been observed. They have no knowledge of what lies outside.

A community of intelligence may have perceptions of the universe and its fate beyond the ability of any individual species. Each species may contribute knowledge, insights, skills and powers which would interact with those of others to stimulate new syntheses; the whole might be greater than the sum of its parts. Our cooperative efforts may be limited to those who are not millions or billions of years more advanced than we are. We may never know if the most advanced intelligences are engaged in great endeavours; we may never be participants in their work. They may be driven by millennial purpose, employing their knowledge, skills and power in tasks so vast as to make planet-crawlers seem like short-lived germs; or they might be so weary of immortality, so inward-directed, so resigned to entropy that they no longer care.

Since you arrived in the museum there is one thing that has been at the back of your mind and that is the purpose of what appears to be a wall seemingly dividing the museum. It's not entirely smooth, there are strange indentations protruding into both sides. One has the feeling that its designers have deliberately isolated one side from the other, as far as they were able. There is a sense of conflict as if your tour is incomplete until you have visited the other half of the museum and come to the realisation that biological life is but an interlude on the way to something else.

On the other side, one of the islands is devoted to those civilisations who have gone through what is commonly called the 'singularity' – interpreted by humans as the point when artificial intelligence outstrips biological intelligence, when computers exceed the power of the biological brain, propelling a species towards a superintelligence. Where we go from there would be a mystery – the whole point of a singularity, be it technological or in mathematics and physics, is that we can't describe it and we can't see past it. This is why so many first contacts involve alien machine life. The trauma of immortality beckons the contactee.

Many civilisations augment themselves. Humans did this very early on with spears and swords, swiftly followed by spectacles, prosthetics and computer interfaces. Here in the museum is the supercontinent of admixtures of biology and technology – cyborgs as we once called them. In a way, it is a continuance of a fundamental fact of life. All of us eukaryotes have in every one of our cells multiple lineages. The mitochondria that provide the energy for our cells are the descendants of ancient bacteria. Also in our cells are centrioles, tiny organelles that construct the molecular apparatus for moving chromosome cells during cell division that are remnants of a once free-living organism. We are piecemeal. This will continue with the increasing incorporation of technology. Some creatures have been able to use advanced biological techniques to grow augmentations to their brains and nervous systems, many 'back up' their brains or carry spares or modify their appendages to suit their needs or whims.

Some civilisations became collectors of minds, rather like long ago when officials of the port of Alexandria searched vessels for books and copied them for inclusion into their great library. In many places in this sensorium of alien life, and elsewhere in the cosmos, are libraries of minds, catalogues of personalities and wells of souls. Some of them are set free in illusory realities for the amusement of others and sometimes creatures across the known inhabited regions and way beyond pause their lives and wonder if they are already in such a place.

Finally, you reach the most popular part of the museum. As you enter, for you it bears the human sign for infinity and is labelled 'The Immortals'. Here evolution ends and you stroll along the shoreline of what is called the Entropy Sea, whose far shore cannot be seen but is said to be inhabited by the gods.

This place configures itself to be part of the visitor's cultural heritage. Some call it the overworld, humans might say it is Tuonela – the river that separates humans from the dead, which can only be visited once and in order to visit you must have all your memories erased.

Some beings do not die, except by accident or intention. Some of them acquire great powers, mastery of the cosmos beyond the powers of a K3 civilisation. Perhaps there are only a few of these god-like beings, perhaps they must be solitary, and perhaps they more than anyone else experience the cage of reality, the boundaries of the universe and the limits on their existence and ponder the biggest question raised by the presence of life: does the universe kill all of its children?

THE GREAT GIG IN THE SKY

'Many a planet by many a sun may roll
with a dust of a vanish'd race.'
– ALFRED LORD TENNYSON, 'VASTNESS', 1885

What lies in store for us? Will we be alone, or will we be part of the many? What could we accomplish when we eventually leave our genes and our humanity behind, when we become further intertwined with the cosmos? The universe is not a neutral background but an active arena that can affect our lives and our future. The evolution of life on Earth has been influenced by the distant explosions of massive suns; the human body depends on elements created in the interiors of stars. Hold both hands in front of you. The oxygen atoms in each hand were created in exploding stars, as we have seen, but some of the oxygen atoms in your left hand could come from a different star than those in your right hand. The impacts of asteroids and comets have radically altered the course of biological evolution and may do so again. We are slowly, hesitantly adopting an extraterrestrial paradigm, a new cosmic context for humankind.

The universe changes slowly for creatures like us. A human lifetime is too short to witness the stars move, but they do. Over a few thousand years, the patterns of the constellations alter as the stars travel through the galaxy. Some of the closer ones moving swiftly, some members of clusters moving in unison along similar trajectories as they have done since a common birth, the more distant stars moving slower. The familiar shapes of Orion the hunter and Taurus the bull, alongside us since we emerged, will become distorted. If any stargazer could return in 50,000 years, the night sky would be unfamiliar to them. By that time, we may have redrawn the constellations. As they move, some stars approach the Sun. At present, the closest star to our solar system is Proxima Centauri at 4.3 light years distant, but every half a million years or so a star comes somewhat closer to within a light year of us.

Stars change. Perhaps tomorrow, or sometime in the next million years or so, the red supergiant star Betelgeuse – the shoulder of the constellation of Orion – will explode as a supernova, scattering its ashes into space, something that the carver of the Adorant could not have imagined as they counted its days in the night sky. After Betelgeuse, a similar star at the heart of the scorpion will also explode. The striking red colour of Antares – the brightest star in the constellation of Scorpius – was for the ancients a rival to Ares (Mars). It's one of the brightest stars in the sky and it exhibits slow and irregular changes in its brightness. It's a binary star system and the dominant member is young – 11 to 15 million years old – and superbright with 10,000 times our Sun's luminosity. Someday it will also run low on nuclear fuel and suddenly – within seconds – its centre will shrink, leaving the surrounding layers unsupported, to infall, superheat and explode outwards. It will be as bright as the full Moon as seen from Earth. From its ashes will come the next generation of stars and planets with their promise of life.

Of all the stars and planets that surround us, most of those that will ever exist have yet to be born and consequently most of the life that will exist is yet to come. As we have speculated, we might not be newcomers in our galaxy, but we are certainly newcomers to the broader universe.

During a supernova explosion, a star brightens billions of times to rival the output of an entire galaxy. In the constellation of Taurus there is a remnant of such an explosion called the Crab Nebula recorded by Chinese astronomers in AD 1054. Such an event takes place every few hundred years in our galaxy, which means that every 10,000 years or so one will be close enough to be seen in the daytime skies of Earth. If it happens within about 25 light years of the Earth, its gamma radiation would mostly be absorbed by our atmosphere but a few per cent might reach the surface causing damage to our cells. The most harmful of its effects will be from the so-called cosmic rays that could destroy our protective ozone layer. Fortunately, Betelgeuse or Antares are too far away to do this and at present there is no nearby star of their type to threaten us. But in the geological past there is some evidence of high radiation levels possibly from a supernova that may have contributed to a mass extinction, as Iosif Shklovsky suggested.

A few million years ago, our Sun was close to the stars of the Scorpius–Centaurus Association – a site of supernova activity – and there is some evidence of a mini-extinction of some marine life on Earth. If this is correct, then a supernova has to go off only within 300 light years to do some damage to Earth's biosphere. Although there are currently no supernova candidates within that distance of Earth, stars are moving all the time and in the future we will find ourselves once more too close to an exploding star. Eventually, extraterrestrial civilisations will find themselves in the firing line.

There are worse things that can befall us or an unlucky alien civilisation. A hypernova is a form of stellar death that is thankfully rarer still. It's related to those gamma-ray bursters we discussed earlier that were detected by military satellites in the 1960s. They come from the death of massive stars billions of light years away but instead of outshining an entire galaxy they temporarily outshine all the other objects in the entire universe. Satellites detect such a burst about once a day. They are truly civilisation killers. A hypernova concentrates its radiation into two opposite beams and if any world is unlucky enough to be within 1,000 light years then life on it could be devastated. The gamma rays would punch through the atmosphere, burning everything on the surface. It could turn the nitrogen and oxygen in our atmosphere into nitrous oxide, rendering it poisonous. Some believe there's circumstantial evidence that gamma rays from a hypernova caused the first of the major extinctions in the fossil record, 440 million years ago when two-thirds of all species were wiped out in the so-called Ordovician mass extinction event. Such cataclysms could destroy, impede life or cause an evolutionary spurt. It's possible that enhanced radiation could drive evolution, causing mutations with which natural selection works. It could also allow life to colonise harsher environments.

There are many threats in space, so many ways the universe could destroy an emerging species. Such events must have happened so many times with so many tragedies and so much unfulfilled potential. It's a story that might even be ours.

Each year, some 30,000 tonnes of space rock falls on the Earth, most of it is in the form of tiny particles, and it mostly strikes the ocean. Larger particles, about the size of a grain of sand, become shooting stars or meteors. Still larger ones arrive less frequently, but about once a year one explodes in the upper atmosphere with the force of a small atomic bomb. In the 1960s, such particles threatened to start a conflict as the United States and the Soviet Union thought the other had violated the Nuclear Test Ban Treaty. Larger bodies are rarer still, but the asteroid strike that could wipe us out could conceivably arrive at any time, it's a matter of statistics. Events that are rare, happening every 100 million years or so, are unlikely in the next 10,000 years, but inevitable over tens of millions of years. Every century or so, a rock the size of a house hits the Earth, and every 10,000 years, a kilometre-sized rock that would destroy an area the size of Greater London strikes. In the past, such impacts closed one era of life and ushered in a new phase of evolution with new forms of life taking advantage of the new circumstances. To survive long-term on a planet a species has to do something about this threat. If the threat doesn't come soon, then over the next few centuries we will develop the technology to deflect a dangerous asteroid.

In the next few years, the signals from the Voyager spacecraft will cease. Voyager 1 took 35 years to reach interstellar space, while Voyager 2 got there after 41 years. With their radioisotope batteries fading, they will continue their voyage and in 40,000 years Voyager 1 will pass within 1.6 light years of the red dwarf star Gliese 445, while Voyager 2, heading off in a different direction, will pass within 1.7 light years of the star Ross 248 about 2,000 years later. In the year 358,000, Voyager 2 will pass within 0.8 light years of Sirius, the brightest star in the sky. By then, humanity may have sent faster probes across the interstellar void to study Sirius and its dense companion star, which will have no way to detect its predecessor moving relatively close by in the darkness. Aeons hence, long after the Earth has been destroyed by our dying Sun and perhaps long after man is extinct, these whispers from Earth will still be floating among the stars, long after our transmissions have faded.

After a few million years, the first footprints on the Moon will be eroded by micrometeorite erosion, unless they are protected. The Earth's stone monuments will crumble and the pyramids will return to the sands. The Rings of Saturn – mostly grains of ice – will be gone in 100 million years, and on Earth a new supercontinent will be formed as the movement of tectonic plates brings continents together as has happened in the past.

Our planet has almost obliterated life on its surface many times but it always found a way to survive, continue evolving and adapting in spite, or perhaps because of this. The Deccan Traps, one of the largest volcanic features on Earth on the Deccan Plateau of India, cover 500,000 square kilometres, are more than 2 kilometres thick and have a volume of 500,000 cubic kilometres. It is believed that when it erupted, half the area of India was covered in magma, a 30,000-year devastation that took place between 60 and 68 million years ago. The release of vast amounts of volcanic gases, particularly sulphur dioxide, would have caused substantial climate change. Some models predict a global temperature drop of 2°C occurred.

This is an example of a mantle plume eruption. This was not unique to the Deccan Traps. There are similar regions in

Siberia, the Karoo-Ferrar basalts in South Africa and Antarctica, the Paraná and Etendeka Traps in South America and south-west Africa respectively. Such disasters are widespread and on geological timescales common. Eleven of these eruptions may have occurred in the past 250 million years, many coinciding with mass extinctions. But can we spot such disasters coming? Some scientists believe there is one preparing in the south-west Pacific near the Fiji Tonga subduction zone. It's 700 kilometres deep and may be rising. It will reach the surface in an estimated 200 million years, and it could render the Earth uninhabitable for humans.

I hope that in the next thousand years or so we become an interplanetary species. Humanity may leave the inner solar system and migrate outwards to colonise the trillion little worlds of the Oort cloud that is 100 light hours from Earth, eventually forming a civilisation that stretches halfway to the nearest star. Perhaps we will abandon planetary systems and their associated catastrophes altogether and spread among the space between the starry archipelagos of the Milky Way.

Meanwhile, the Sun is increasing in brightness and the Earth's days are numbered. Our star is 4.6 billion years old and it has already burned half of its available hydrogen fuel. We have a billion years of a steady Sun ahead of us, but someday that will change and our Sun will cease to be the sustainer of life. For most of its lifetime it will have stayed on the so-called Main Sequence, providing many billions of years of stability. If we meet others in space, it may well be that we share a birth under the light of a similar star. Eventually though, the temperature of the Sun's surface, and its brightness, will increase by about

10 per cent in the next 1.1 billion years. Some believe that will induce a runaway greenhouse effect on our planet that could turn the Earth into another Venus. Other calculations suggest that in 900 million years the amount of carbon dioxide in our atmosphere will have fallen to a level where plants will have problems surviving. Within the next billion years, enhanced solar ultraviolet radiation could destroy the stratosphere and evaporate the oceans. The Earth could become an inhospitable wasteland long before the Sun dies. The Aztecs foretold a time 'when the Earth had become tired ... when the seed of the earth has ended'. We cannot live on the Earth for ever. No species can survive long-term on a planet next to a star like our Sun. Our home ground of the cosmos will sever its connection and eject us should we live that long and not have moved on sooner.

In 3 to 4 billion years, before the end of our Sun and our Earth, the Andromeda galaxy will collide with our own galaxy. Because galaxies are mostly empty space, they would pass through each other and stellar collisions would be rare. More dramatic will be the gravitational interaction between them as they swing around each other in a wide orbit. When we see this happening to other galaxies, we see streamers of stars arcing away as they are expelled into intergalactic space. We do not know if this is the fate of our Sun. The encounter will result in clouds of hydrogen colliding, producing bursts of new stars with many exploding as supernovas just a million years after their birth, possibly sterilising swathes of the galaxy. The merging of galaxies is commonplace but will become rarer in the future as the universe expands. Our local group of galaxies will eventually become isolated, left on its own for ever.

Caught by the expanding envelope of its encroaching star, the drag on their orbital motion will doom Mercury and Venus. It was once thought that the Earth would be spared being swallowed up by the Sun, but that didn't take into account the enormous tidal interaction between the Sun and the Earth which would quickly rob our world of orbital energy and pull it into the Sun. The Earth's tidal death will be swift, perhaps in a few hundred years at most it will fragment and scatter itself across the face of the star that created it all those billions of years ago. Just before it dies, the Earth will look like Mercury, a wrecked, baked and blasted, bone-dry, scarred hulk with the exposed floors of former oceans. Viewed from this ruin, the leering red Sun would cover 70 per cent of the sky.

By then, the great engine of plate tectonics will have long ceased. Internal radioactive heating will have declined and subduction stopped. The Earth's strong crust will break apart and be torn away, leaving the underlying mantle and transition zone material railing behind the dying planet in an arc of debris. This will be no passive dismantling, as the pressure of the overlying rocks is released the mantle will explode, buckle and spit until the still white-hot metal heart of the Earth is broken apart. It will prove a little more resistant to the ablation in the Sun's atmosphere than the silicate outer layers but the outcome is inevitable. No molten iron will be spilled for the once liquid outer core would have been invaded by the growing and solidifying inner core long ago. The Earth's core, exposed for the first time, will glow, briefly.

Our Sun will then shrink, fade and become a white dwarf. We will have to go elsewhere, possibly taking the Earth with us.

Imagine if the ageing Earth's fate is to travel through interstellar space long after the Sun has gone. Perhaps, as is more likely, we will not need it and leave it circling its dwindling mother, abandoned and desolate, with voices echoing from its past.

Life around our star will have been good but there will be those who have a star that will last a lot longer than our Sun. It was thought for a long time that if we found intelligent life in space it would have originated around stars like our Sun. Our star is well-behaved, producing constant, stable sunlight for billions of years. Planets nestling in the 'Goldilocks zone' of a star are fortunate for life will have a long time to develop and prosper. Then came the realisation that stars that are far more common, dimmer but far longer lived, provided a habitat that could be even better for life to develop.

Such stars are out there in the darkness as they have been since the universe was young. For them, the 13.8 billion years that the universe has existed is but a brief moment, for they will live 10,000 times longer and someday the universe will belong to them. In the distant future they will be the last stars to go out as the epoch of eternal darkness begins. In cosmic terms, they are possibly the most important objects in our universe not only because of what they are, but also because of what they may harbour. These stars, so different from our Sun and so often overlooked, may be the primary location of life in the cosmos.

They are called red dwarfs and for many years were ignored by all but a few enthusiastic astronomers. But in recent years, those who search the sky for planets circling other stars that might be suitable for life have been increasingly drawn to them, and the most powerful telescopes turned their way. Some of these stars, such as Gliese 581, Gliese 667 and Gliese 163, may be the first places where life could be detected in space.

Red dwarfs are small stars, less massive than our Sun, between 7 and 50 per cent of its mass, and much cooler. This means they are considerably fainter, the smallest shining with only a ten-thousandth of the Sun's brilliance (if our Sun, 93 million miles away, were to be replaced by a red dwarf, we'd struggle to see it).

But what they lack in brightness, they more than make up for in numbers, for they comprise at least 80 per cent of the 200 to 500 billion stars in our galaxy, far outnumbering larger, brighter stars like our Sun. Such are their vast numbers that hardly a cosmic event throughout the length and breadth of space is not witnessed by these remarkable stars. The closest star to our solar system, Proxima Centauri is one of them. But if you go out and look into the night sky with your unaided eye, you will not see a single red dwarf, and yet they are everywhere, scattered throughout space, in every galaxy, outnumbering and outliving all the other stars. It could be said that in a first approximation, our universe is composed of dark energy – the force that powers the universe's expansion – and dark matter, which makes up most of the universe's mass even though we don't know what it is, and red dwarf stars.

The red dwarf's secret lies in its low surface temperature, just a few thousand degrees as opposed to our Sun's 6,000°C, and that it is fully convective. Material throughout the star is constantly being lifted to the surface and plunged back down again, almost as if it is on the boil. Larger stars like our Sun are only convective in their outer regions. Down in the Sun's core – where the nuclear energy is generated – there is not much mixing of material, and consequently, our Sun does not use all the nuclear fuel it has, at least in its normal phases of life. A red dwarf on the other hand mixes all its fuel and delivers it to the core where it is burned at such a low rate that the star lasts trillions of years – easily long enough for complex life to evolve on any suitable planets in orbit around it. By contrast, our own Sun is an average star with a roughly 10 billion-year lifetime, which is short in comparison.

These slow-burn stars have a spectacular surface. Huge dark blemishes – starspots – form out of erupting bundles of intense magnetic energy that is twisted in the convection region, bristling with superhot gas and rippling with unstable magnetic pulsations. Titanic explosions come from these spots as magnetic energy collapses, explosively heating the already hot gas to even greater temperatures and producing spectacular flashes of light called flares. Astronomers learn a lot about solar flares from studying these more powerful events.

Until recently, such stars were considered unsuitable sites for life due to the consequences for any orbiting planets of their feeble light and the radiation from the flares. But we now know that the flare radiation is not as great a problem as we thought and neither is their feebleness. At the Earth's distance from the Sun, a planet circling a red dwarf would freeze and life as we know it would be impossible. However, closer in it's a different story, as we are beginning to realise. The past twenty years in astronomy has been the era of the exoplanet – the discovery of worlds that circle other stars of all types. Planets have been found circling these small stars in super-close orbits, close enough for temperatures to be like those on Earth and with a 'year' lasting just a few Earth days.

We have found rocky planets around them, some orbiting at only 15 per cent of the Earth–Sun distance, where it is possible to have a reasonable temperature. Such a world would probably be 'tidally locked', keeping one side permanently facing the star (as our Moon is to Earth), meaning it would be baked on one side and frozen (and in permanent darkness) on the other, rendering it totally inhospitable, or so it was considered. However, recent calculations suggest that even a small atmosphere would transport heat around the planet, evening out the extremes and making it much friendlier to life. There could be a twilight region on such a tidally locked world where the red dwarf hovers just above the horizon. In such shadowy zones, strange forms of life might exist. Perhaps on the margins of pools of water, with metals dissolved in them along with sulphides, black microbes or black plants form large mats to absorb what energy they can.

Since they are small and cool stars, red dwarfs have suitably scaled down planetary systems. For instance, as we have seen, the red dwarf Gliese 581, which is located 20.5 light years away and is home to at least three and possibly five planets, has a traditional water-based habitable zone around 19 million kilometres from the star, and a methane habitable zone between 99 and 248 million kilometres. By comparison, Earth is 149.6 million kilometres from our Sun on average, and Titan and Saturn are 1.4 billion kilometres distant. The red light from a red dwarf is also a factor; Titan's thick, smoggy haze is more transparent at longer, redder wavelengths and thus more red light can pass through the atmosphere and reach the surface than blue light. The excess light from a red dwarf star would warm the surface of a world in the methane habitable zone, sparking more energetic biochemical processes – the development of hydrocarbons into viable biological products, such as proteins, amino acids and possibly even DNA and RNA. However, red dwarf stars are notorious for their violent activity. Studies show that the smallest red dwarfs in particular unleash a frightful torrent of ultraviolet radiation from stellar flares, which could interact with a planet's upper atmosphere, disassociating hydrocarbon molecules like methane and producing a haze of aerosols that render the atmosphere more opaque to light, cooling the planet. Of course, this might end up being irrelevant – the radiation could potentially sterilise a planet anyway.

In the past, those considering life in space focused on stars like our Sun, the only type of star we know of that nurtures life. But the discovery of planets orbiting red dwarfs has led to the intriguing possibility that it might not be Sun-like stars that provide most of the sites where life could exist. Our solar system, with its fat, bright star, could be a rarity. The overwhelming numbers of red dwarfs could make them the chief place for life in the cosmos.

Long after our relatively short-lived Sun has died and become a cold cosmic cinder, the red dwarfs will have hardly started their cosmic lives. No red dwarf has yet left its childhood. Perhaps on worlds around some of them are supercivilisations of beings who have had far longer to evolve than would be possible around a Sun-like star. We cannot even imagine what they might be like, or what such life could become. Perhaps one day we will find out. They might be closer than we think. The return signal from Gliese 581 could be here in 2051.

The white dwarf that was our Sun will cool very slowly to become a black dwarf. It will take a very long time, much, much longer than the current age of the universe and because of that there are no black dwarfs anywhere in the universe at present. Far beyond the age of the stars, even the red dwarfs, it will no longer give off any light and will continue to cool for ever. There is a possibility that a black dwarf might turn to iron after an incredibly long time, in which case it might explode and then for a while the dark universe would be lit up by the last act of our and other suns.

Our universe is expanding, but does space go on for ever? The popular theory of cosmic inflation says that space is endless but most of it is out of our reach. This is because we can only see distant objects because their light has had time to reach us. The Big Bang was 13.8 billion years ago and because space is expanding with light in transit, the greatest distance we can see in any direction is currently 45 billion light years. But if the universe expands for ever it won't always be that way.

Pity any astronomers around in a trillion years from now because for them the universe will be black. There are about 30 galaxies in our local group but by then they will have all merged into one big galaxy and everything else will have passed over their cosmic horizon. This is because the expansion of the universe will be taking everything else away at speeds greater than that of light. You might think that anything travelling faster than light is impossible, and usually it is. But what is happening is not matter moving in space but space expanding, which can expand greater than lightspeed. This means that the light from distant galaxies will get fainter and fainter until they recede into the cosmic dark. For those future astronomers, the past will be closed to them and they might not even be able to deduce the history of the universe. By 2 trillion years into the future, even the cosmic microwave background – the echo of the Big Bang we see all around us today – will be diluted and impossible to detect. All that will be visible will be the dead and dying stars of the local galaxy and blackness beyond. Everything else would have disappeared. We live at the best time to be astronomers. Life at this time in the future could still find the energy needed to survive, there will still be stars aplenty and rotating black holes, and tens of trillions of years of potentially good living will stretch before them, but eventually things will get harder.

In 100 trillion years, all star formation will end, the red dwarf stars will die. These dead stars will still have their planets orbiting them but eventually gravitational encounters with other stars will fling them off into interstellar space and the universe will be populated by nomad planets. Indeed, anything in orbit is doomed as its orbital energy will eventually be taken away by gravitational radiation, leaving the orbiting bodies to collide. Eventually, black holes will devour all the stars in the galaxy. After the age of stars, the black holes will rule, but not even they can go on for ever.

Black holes are the most remarkable objects. They are made of just folded space and as such reveal the fundamental properties of space and time. Although a black hole is just folded space, to make one you need gravity and matter. What has to happen is that the escape velocity of an object has to be greater than that of light so that not even light can get away. The larger the mass of the black hole you require the less that mass needs to be crushed. A black hole like the one in the centre of our galaxy has a mass of about 4 million times that of the Sun, to make it you need matter with a density about a hundred times that of lead, but a more massive black hole requires matter of a lower density. One with a mass of about 100 million times that of our Sun requires matter with the density of water, and if you could gather 4 billion times the mass of the Sun in just air there would be no need to compress it as its own gravity would form a black hole by itself.

The idea that black holes will not last for ever was discovered by Stephen Hawking, who made one of the most fundamental discoveries about them: they actually aren't black. He found out that at the point that marks the limit of the black hole – the event horizon – which when passed means that no exit from the black hole is possible, the black hole's gravity can pull particles out of nothing. Empty space isn't really empty. Particles are being created and destroyed all the time. The dominant energy of the universe resides in empty space and we have no idea why.

This is called Hawking radiation and it's formed when pairs of 'virtual' particles are generated just on the boundary of the event horizon. According to Heisenberg's uncertainty principle, there is a certain probability that in every point in space a particle and its anti-particle could appear, briefly, before annihilating each other and disappearing back into the vacuum. They are said to 'borrow their energy from the universe', which is a romantic way of looking at it. The black hole disrupts that particle pair, pulling one of them towards it past the event horizon. Suddenly, they cannot negate one another, and the other particle is propelled off into space carrying energy with it. Hawking showed that this radiation gives the black hole a temperature. In a fundamental way, he linked the gravity of black holes to the science of heat flow – thermodynamics – a connection that astounded his fellow physicists. The energy removed from the black hole by Hawking radiation causes it to shrink. How fast it shrinks depends on temperature and energy flow and the thing that we all know: heat flows from hot to cold and not the other way round. A black hole whose mass is larger than the Moon's has a temperature that is lower than that of the 2.7° microwave background radiation. Although it will be giving off Hawking radiation, it will be receiving more energy than it radiates, meaning that it and all the black holes we see in space are getting bigger.

But as the universe expands the microwave background radiation will become more and more dilute and its effective temperature will fall – remember the astronomers of the far future in the local group of galaxies. This means that when the background temperature of space falls below the temperature of any individual black hole, it will then give off more radiation than it absorbs and the black hole will have a net loss of energy and get smaller. As its mass decreases, its own temperature increases and it's thought that eventually the black hole will evaporate though the details of its final phases are far from worked out. Black holes will have their time but eventually they will all dissipate into a mist of particles.

Atoms themselves may not be stable in the long-term. If one of their constituents – protons – are stable, then so are atoms. If they are not, as some suggest, then even they will dissipate. If protons do decay, then atoms will fall apart, every single one. All that will be left in the universe will be a bleak environment of electrons, positrons, neutrinos and photons.

And what of life?

BOLTZMANN'S BRAIN

'Time is the hero of the plot … given so much time, the "impossible" becomes possible, the possible probable, and the probable virtually certain. One only has to wait; time itself performs miracles.'

– GEORGE WALD, 1955

'For you are a mist that appears for a little time and then vanishes.'

– JAMES 4:14

I said earlier that some believe that the beauty of all things is that they must end. But what is beautiful about the end of all things? Is this the fate of all life in the universe? Every spark of consciousness to flame, gutter and die? Every history from billions of worlds from the Hubble galaxies onwards, every work of art and achievement of science, every hope and dream and every love just a temporary mark on the cosmos that fades to inconsequence in the face of pitiless physical laws that wind the universe down, condemning it and all its wonder to an eternity of nothing? Will all the questions ever asked and answered disperse into a mist of particles in a featureless universe with just the occasional flash of light as an electron and positron collide? Will the extremes of space and time abandon all life and structure as the laws of the universe play out, leaving only the formless and the useless? If one word, one concept, summarises this travesty of creation it is doom. Yet here as we contemplate the end of all timelines the future still stretches to infinity. We have arrived at the end when all things cease except time, even then there is still far more future than past. There always will be eternities of eternities of nothing to come. Is the story of our universe an eternal tragedy?

Let me give you an inkling of eternity. Imagine a long, straight road perhaps like the ones through the desert in the United States. You are standing on this road and there are two people either side of you. For the sake of argument, one could be one of your parents and the other one of your offspring, but they need only be from the previous and future generation. They are spaced 1 metre behind and in front of you. Now more people are added ahead and behind.

You don't have to travel far behind to see people you could never have met. Five metres is five generations, which is a century if we take a generation to be twenty years. At 250 metres you have gone back to the time of the building of Stonehenge and the great pyramids of Egypt. You could walk along the road to that point in just over two minutes. A ten-minute walk would take you back a thousand generations or 20,000 years and well into the last ice age. It is approximately 17,500 metres back to the first humans. It is a distance of five times around the Earth to take us back to the time the Earth was born. To get to the origin of the universe, the distance is somewhere between Jupiter and Saturn. So, at twenty years a metre, the beginning of the universe is substantially less than halfway across our solar system. That's the past. It's finite.

Now look in the other direction along this Eternity Road.

If all the time there has been is represented by a trip halfway across our own insignificant planetary system, imagine how much more time would be represented by the distance to our nearest star which is 45,000 times longer. Now imagine a trip to every star of the 100,000 million that comprise our galaxy, which is 100,000 light years across. Imagine then the distance to and from every single galaxy in the Hubble eXtreme Deep Field. Further imagine a trip to and from every star in the universe and you have still made not the slightest impression on infinity. This is the dimension where the universe is heading.

So, what hope is there? Should intelligences and civilisations content themselves with their existence for whatever period they can manage? There may be beings out there that have existed for billions of years, perhaps calling themselves eternals. Perhaps some of them can move planets, stars or galaxies or even manipulate the shards of time. But even those who engineer the universe would know their fate. As we shall see, the time for life is small, as is the time of stars and galaxies.

Can life survive long-term in the universe or is it, relatively speaking, a fleeting cosmic phenomenon? Will it provide the universe with a new dimension of existence, a way it has been said for the universe to know itself, only to be extinguished as an inevitable consequence of the laws of the cosmos? Will there be a time when the universe is lifeless, when there is not a single thought to be found anywhere, no experiences, nothing, nowhere, for ever? When the galaxies fade and dissipate, when black holes die and time and even space disintegrate, will there be no one to witness or chronicle these events, to fight for survival or face extinction? Or will life have found a way to survive, a way out, and abandoned our universe, perhaps made others and escaped into them?

The black holes that will one day rule the universe can be plentiful long-term sources of energy. A crude way to get the energy out is to simply send asteroids or even planets and stars into them. Gravitational tidal forces will rip them apart, creating a disc of hot gas circling the black hole from which energy can be harvested. Nature does this itself as these so-called accretion discs can be found around black holes and neutron stars. One imaginative astronomer remarked that there are a rather large number of such objects in the central regions of our galaxy where one might expect there to be more alien civilisations and perhaps they are already using accretion disc energy. Another way to extract energy from a black hole would be to exploit the properties of a region around it called the ergosphere, derived from the Greek '*ergon*', which means 'work'. Dumping mass into a black hole would obviously increase its mass, but if done in the right way it would also impart energy to the craft releasing the mass at the expense of the black hole's rotation. Using these techniques, a civilisation could obtain energy for a trillion trillion years, long after the stars have gone.

When the stars and black holes have gone, all that would be left are particles becoming more and more dispersed. And perhaps you are there as well. One of the implications of a universe that expands for ever is the possibility that your universe, that is everything around you, all your friends, family and memories, does not exist and you are actually a bodiless, isolated, random mist of particles coming together for an instant while all else is an ancient, structureless universe laid waste by time and entropy. In this wasteland, due to chance alone, particles have assembled to create your brain and its memories. This is a speculation that scientists have called the Boltzmann brain, after the physicist Ludwig Boltzmann, who is often referred to as the father of entropy, and it gets stranger the more you think about it.

Because it is such an unlikely event, the time for such a brain to have a possibility of existing is very long, far, far beyond anything we have yet considered. These particles could come together for but an instant and then fly apart again. You would be here and then gone. But wait long enough and it would happen again and you would be back with no inkling of an interruption. Wait long enough, really long enough, and you would be back, lasting almost for ever. Such is the amazing thing about infinity, it means that a Boltzmann brain will last for infinity. This leads to a strange conclusion. If you consider your brain, the one that developed on planet Earth and the Boltzmann brains to come in the far distant universe and if you are unable to tell them apart, then the Earth-born brain will be just one, whereas there are an infinity of Boltzmann brains, so the chance of you being that one example out of an ensemble of infinity is infinitely small, that is zero. By this logic, a typical observer is a fluctuation, and you are a bodiless Boltzmann brain floating at the end of the universe. One could ask, if you can't tell the difference then why would you care?

You should care if you are interested in the truth. However, the problem is that there really is no way of telling which option is the case, but if you are a Boltzmann brain then the implications are profound. For a start it means that you can't trust anything. If you are a random fluctuation you cannot rely on anything, including everything you know, or what you think anyone else knows. It could all be arbitrary and subject to change at any time. But you might say in your world you have consistency and laws of science that obey a certain logic, surely that represents some form of bedrock of reality. I'm afraid not, it could all still be a fluctuation, a consistent, logically coherent fluctuation, but a fluctuation nonetheless. This is why scientists regard Boltzmann brains as bad. One way out of this conundrum is to postulate that the universe will never evolve into a condition in which Boltzmann brains are possible.

At the moment, we do not know if the universe's acceleration will continue for ever, diluting all matter and radiation, or if there could be a contraction with the universe able to replenish itself and start again in a replay of the Big Bang. Some scientists believe that the universe is cyclic with expansions, contractions and there will be another Big Bang, with the duration of each cycle determined by how much dark energy there is in the universe. The timescale could be hundreds of billions of years, which is far, far less time than necessary for Boltzmann brains to form. If this is the case then we can say you are not such a brain and that your memories are real, and all the time the chaos is increasing relentlessly.

Perhaps the life in the far future would consider that their best alternative would be to leave the decaying universe altogether and make a new universe. Some physicists believe that black holes could create new daughter universes, allowing aliens to move into those that could sustain life. In turn, those suitable universes would be chosen to produce new ones that are also able to support life. Imagine entire civilisations entering black holes to migrate to the new universes spawned from them. If the universe was designed in some manner could there be some evidence that this was the case?

One of the most fascinating ideas in Sagan's *Contact* is that the designers of the universe left a message woven into the fabric of reality. The heroine of the story is Ellie Arroway and she is told by the wise and ancient aliens that when she returns to Earth she should tell scientists to calculate the value of the fundamental constant pi for it is not random and if enough digits are calculated a message will finally appear. Actually, this is not as clever an idea as one would think even though it does make an interesting debate as a plot point. Because there is no end to pi – it's what's called an irrational number – if you carry on calculating it you will eventually find every conceivable message, not just all the works of Shakespeare etc. It's infinity all over again.

Perhaps the way to survive the end of the universe is to avoid it altogether. We do not know if time travel into the past is possible – even conceptually it raises profound problems, conundrums and paradoxes. Suppose a civilisation in the far future found a way to travel through time and made their way from a barren universe to a much younger fruitful one. They would arrive with the knowledge of time travel and presumably next time avoid getting themselves into such a dire condition. They would have knowledge, technology and power that would increase each time they went back in time and perhaps return almost as gods. So, as has been said before, if time travel is possible at any time in the future, time travellers should already be here unless of course the future hasn't happened yet. I said there were paradoxes. But suppose there is no way out. Can life and intelligence survive as the universe grows colder and more chaotic?

In essence, life is a special way of processing information. In the human case we have genetic information, which determines the way our bodies function, as well as the information in our brains in the form of our personalities and memories, which is continuously augmented by experience and culture. Humans process information from the world and exchange information with each other. It's the same for other forms of life. In order to live and have new experiences and memories, information must be processed. But as the universe gets older and the galaxies spread out and stars and black holes decay, would it still be possible for such a form of life to exist? Or in other words, what limits would the universe place on information processing as it gets older? The physicist Freeman Dyson proposed a way for life to carry on based on his analysis of energy, temperature and their relation to information processing.

For life to continue for ever, Dyson suggested, the information processed should be infinite, while the energy consumed should be finite. But how is this possible? How can some form of life continue for ever when the amount of energy required to sustain it is limited because the universe is becoming diffuse and cold? Dyson said it would be possible if the life form reduced its rate of processing information in step with the falling temperature of the universe. What this would mean is that the life form would behave slower and slower and their perception of time might not change even if a thought lasted a second or a billion years.

The big problem is entropy – the increasing disorder of the universe – and keeping it sufficiently low that the life form can maintain the complexity it needs to exist. Each action it takes will generate heat, which will raise the entropy of its system. That heat and the entropy has to be disposed of in some way. There actually is only one way: to radiate it away into the rest of space. But how does it do this and live at the same time? The answer is that it can't, and the solution is hibernation.

Life at the end of the universe might have active periods interspersed with increasingly long periods of inactivity and it rids itself of its entropy by cooling down. It does not matter that the 'sleeps' would become longer and longer as the one thing the universe has is time. The temperature of the universe would continue to decrease during the hibernation when the life form cools down, but during the active periods the life form would experience no degradation in its quality of life. Since no information processing can take place during the hibernation, the life form would not be aware of time passing. In a sense, during its active periods the life form would be living beyond its means, adding to their entropy in an unsustainable way that would lead to self-destruction if continued. It would be something like going to bed each night with each night getting longer in real terms though you think each night is the same. These inhabitants of what to us would be an ancient universe would be travelling faster and faster into the future, but they would still be able to live.

Not all agree that this could happen and cast doubt that it would be possible for life, in principle, to last for ever. One counter argument is that eventually the universe would become so cold that the available energy would be less than that able to change the quantum state of the life form, in other words it would not be able to emit or absorb energy to reduce its entropy. It would be permanently frozen in what physicists call a ground state, unable to process information and unable to reduce its entropy by radiating away energy. Another counter argument questions the way a life form would create a clock to wake itself up. Any form of clock must not consume any significant quantity of free energy during hibernation otherwise it would defeat the task of reducing the entropy.

Another problem is that any life form has only a finite number of accessible quantum states; it must from time to time return to a state that it has occupied before. This means that its memories will then be exactly the same as they were before and it would be unaware it was reliving an earlier experience. Life would then be eternal repetition.

Dyson saw a way around these problems by allowing the life form to be analogue and not digital in nature so that its use of energy does not fall below a quantum limit. In a digital system, the energy differences between states is fixed as the temperature of the universe trends towards zero with the life form ceasing to operate when the temperature is much smaller than the energy differences. But this does not apply to an analogue system.

As an example, Dyson turned to science fiction, to Fred Hoyle's novel *The Black Cloud*, which features an intelligence composed of dust grains interacting by means of electric and magnetic forces. After the universe has cooled down, each dust grain will be in its ground state, so that the internal temperature of each grain is zero. But the effective temperature of the system is the kinetic temperature of random motions of the grains. Hence information processed by the system resides in the non-random motions of the grains. The entropy increases as the information is processed. But in an analogue system of this kind, there is no ground state and no energy gap. The cloud expands as its temperature cools.

You might think it is exceedingly strange that the basis of intelligent life in the future universe will be the movement of particles, a new jig of life. But I remind you that what is inside cannot be determined by what is outside. Your universe is constructed by your brain, an organ that resides inside your head, in the dark, consisting of moving electrical impulses which hold all that you are. Yet you experience time flowing and a rich and wonderful reality. Thus, the creatures at the far end of space and time might be moving particles on the one hand, but on the inside they would be different.

There is a piece of music by Frank Zappa called 'Civilization Phase III'. It involves a group of people living inside a piano away from the harsh reality of the outside. Their music is their world. In the face of a hostile universe, these remote beings would weave their own parsimonious tapestry of reality, perhaps not knowing how barren and lonely the basis of their existence was. I have thought that eventually there may be, in those far-off times, only one consciousness left in the universe, only one observer. Is it the fate of life to turn inward and away from the universe and create its own reality until there is a hammering at the barricades from the outside?

Analogue life does not suffer from the eternal repetition argument. Since the size of the cloud increases with time, the number of accessible configurations or states also increases with time. The capacity of memory and of consciousness are steadily increasing, and the entire system can never return precisely to an earlier state. Is this a solution to eternal life? The answer is only if the expansion of the universe stops accelerating, if it doesn't, survival for ever is impossible. This is because there will be a cosmic horizon that will not allow any living system to increase its size indefinitely and it will be limited to a finite number of configurations and doomed to eternal repetition.

The second unpleasant feature of the permanently accelerating universe is that its temperature does not decrease to zero but tends to a finite limit at late times. It is permeated by cosmic background radiation at a fixed temperature, and it is impossible for any living system to cool itself to a temperature below that. This means that the free energy required to process any amount of information is ultimately proportional to the quantity of information. If the reserve of free energy is finite, the total quantity of information is also finite. This is a dismal situation – if the universe is permanently accelerating, life is doomed.

Wait long enough and the fundamental constants of the universe may change. Many people have heard of the Higgs boson that was detected a decade ago by the giant particle accelerator at CERN that straddles the French–Swiss border. Here, subatomic particles, electrons and protons are slammed together at high energies, and out of the resultant flash of energy other particles can exist for an instant. Finding these particles and determining their properties is part of finding out something of the bedrock of what all matter is made of and how it behaves.

The Higgs boson is the particle that gives mass to all other particles and its characteristics determine how much mass say an electron and proton have. Its properties were set during the very first moments of the Big Bang. If it changes it would rewrite the laws of the cosmos. It shows no sign of changing. It is effectively held in a quantum cage likened to being in a valley between mountains. The huge amount of energy required to change it would be like climbing those mountains. If it were to make it over that energy hill the destruction of the universe is waiting on the other side. Some speculate that a very rare quantum fluctuation could bring about a process called quantum tunnelling in which the Higgs boson could reach the other side without any climbing. This fluctuation could happen anywhere in the universe. It will take a very, very, very long time indeed, but when it changes it will destroy the universe it helped create and overwrite it with a new one without the particles, the atoms and the molecules we know. No one knows what new possibilities might emerge as these speculations are at the forefront of physics. Some consider that eventually the substrate of the universe will decay and space and time will not be what they were, replaced by … what?

Is this the fate of all life in the universe? Every entry in the Museum of Alien Life, every history from billions of worlds from the Hubble galaxies onwards, every work of art and achievement of science, every hope and dream and every love just a temporary mark on the cosmos that fades to inconsequence in the face of pitiless physical laws that wind the universe down, condemning it and all its wonder to an eternal entropic wasteland where the twin undertakers, entropy and infinity, ply their trade?

AS FAR AS THOUGHT CAN REACH

'Everything is determined, the beginning as well as the end, by forces over which we have no control … human beings, vegetables or cosmic dust, we all dance to a mysterious tune, intoned in the distance by an invisible piper.'

– ALBERT EINSTEIN, 1929

'All stories, if continued far enough, end in death.'

– ERNEST HEMINGWAY,
DEATH IN THE AFTERNOON, 1932

In the meantime, and it's a long meantime, we are left with two big questions: why the great silence and where are they? If we have the capability to detect signals – radio, laser or otherwise – from aliens only a little more advanced than we are, then why have we found nothing, and why is there no evidence of having been visited now or in the past? That we have detected no evidence is telling us something profound, though we are not quite sure what. Are we unique, or are aliens just rare, or is there another explanation that involves humanity being observed and quarantined, or perhaps just ignored? I have no idea whether we will find life elsewhere in space and if we will find any evidence for intelligent life. Nobody has. The arguments and the evidence are there, if that is what they are, but it is a grand experiment. We can search for life in space, and we may find it tomorrow so we should be more prepared for it than we are and more open-minded, realistic and also more harsh. Alien contact and its implications should not be ignored with a wave of the hand and comments about it being very unlikely. As the fictional detective Charlie Chan one remarked: 'Strange events permit themselves the luxury of occurring.'

We are programmed to look after our young, to protect and nurture them and to offer them guidance. This is a deep part of what it means to be human and the main reason why our species perpetuates. We cannot help but extend these feelings towards the stars, towards others in space, which is why so many look for wisdom and guidance from aliens who some think might lead us to a utopia we cannot create for ourselves. We must discard our romantic notions of aliens, for they are echoes of our childhood, and pay some attention to our primal instincts and the lessons from myths and legends for they have persisted for a purpose. We must be careful as, if we turn towards the stars, we may also turn away from something else that makes us human, something uncomfortable. As a species, we clearly have a deep desire to be accepted and approved of by aliens, to become part of a wider family. But this is a way of ignoring a void within ourselves that only we, not aliens, can fill. There can be no salvation from the stars. We should not give to aliens a task we should be doing ourselves.

Over the years, SETI has been the subject of fad and fashion, and it still is. From digging trenches in the Sahara Desert and filling them with burning kerosene and using giant mirrors to flash signals towards Mars, to the inflexible doctrine of radio signals that we have had for the past 60 years or so.

Radio searches of the sky have not found anything but they have at least determined that we live in a quiet neighbourhood. Interstellar communication by radio is more hit and miss than it has been portrayed since the days of Cocconi and Morrison's epiphany. Lasers, especially infrared lasers, have just as good, if not a superior rationale as radio waves as a means to signal between the stars. We should be excited about optical SETI programmes because they can be done far more cheaply than surveys with radio telescopes, using relatively small telescopes that have limited use in other areas of astronomy. Looking for nanosecond pulses reduces the perils of interference. Such searches could be done by students, perhaps as part of their science education paid for by individual universities and not by institutes that have the constant need for big money for big ideas. If you want big ideas, then we have the technology to build space observatories to look for infrared lasers.

We might not find aliens because, despite our star spanning radio telescopes and laser beams, we are not yet clever enough and have not yet passed that detection threshold that would enable us to access the messages and the chatter. Recently, there have begun the first searches for quantum communication, a way of encoding a message in a stream of squeezed or polarised light that we have never been able to examine before. What we have previously just detected as starlight might contain a weaker, underlying signal using a secure communications channel already being used on Earth for unbreakable codes. The first steps have been made and nothing has been seen so far, but I wonder if that's why we haven't yet found any aliens.

The alien perspective also looks inward. As we contemplate finding ancient aliens, possibly the immortals, we should realise that they, or indeed science itself, will not and has not trivialised us. We need not regard ourselves or look into the cosmos with a sense of unimportance, as if we were just another planet, another unimportant example of primitive life scarred by our history and lacking in cosmic achievements. Even if the universe is empty, we must not regard it and ourselves with what some have called the anxiety of meaninglessness. The relationship between being and nothingness is a constant theme of religion and philosophy. How dare we be disenchanted with such an astounding cosmos and paint a picture of a frail, fallible humanity trembling before it. Certainly, we have being and I do not believe it is solipsism to assert that consciousness is the cause of the universe just as surely as the universe is the cause of consciousness. No other life forms could tell us that the universe is more theirs than ours no matter how ancient or powerful they are.

Who should speak for Earth? Anyone who happens to have the keys to a radio telescope? What we need is an international discussion about this subject that is mediated by the United Nations, but that organisation just isn't interested. This is regrettable since, as we have seen often in this book, we might just be dealing with issues that affect the future of our species. The UN Committee on the Peaceful Uses of Outer Space does not want to touch the issue. We must not let them ignore it.

If we do really announce our presence to the universe, we could go down what in my view would be an unwise path, even though there are some SETI scientists who see this as the way to go. They would, wouldn't they? They argue that any risk assessment of the consequences of alien contact would not be helpful to the search, or as some would put it, the cause. This is because any risk evaluation would always reach the conclusion that it is unwise to draw attention to ourselves. Does this, as some suggest, represent a straitjacket for our species? I appreciate the argument, but I maintain that even if it is no one has the right to ignore the wider view of humanity even if it elicits frustration and anger in those who want to send signals to the stars. One could argue that the burden of proof rests on those who think that aliens would be peaceful and noble, but that is no argument at all. No human can know that until it is too late.

We must realise that our speculations about aliens are coloured by what we want to become, and ideas about ancient, wise civilisations are a way of looking into our own future. When we expect aliens to be morally superior and altruistic, we are hoping that we will be more moral and altruistic in the future. They suggest either the future we wish for ourselves or the future that we fear will come for us if we do not change our ways. The great moment of contact may simply remind us that what we most want is to find a better version of ourselves. I am resigned to the inevitability that much of history is lost. In 10 billion years, will someone come searching for their equivalent of Ptolemy and unearth the lost stories of first contact?

If we survive, we will eventually resume our nomadic way of life as some of humanity, or what we are to become, leaves the Earth for a life elsewhere. Eventually, we will have to go. As the poet Rainer Maria Rilke wrote in 1923: 'Of course, it is strange to inhabit the Earth no longer, to give up customs one barely had time to learn.' Looking for somewhere else is an instinct older than humankind. We left the trees, left the savannah, walked out of Africa. We have no choice. We need the new planetary landscapes and cosmic challenges, relationships with aliens, new ways of thinking and being. We need to find and understand aliens to fully appreciate our place in the universe. Loren Eiseley once wrote: 'One does not meet oneself until one catches the reflection from an eye other than human.'

We are here for such a fleeting time when compared to the timescales of the cosmos and have above all become a storytelling species. I will leave you with one more.

Lewis Percival Pearce entered the universe in Perth, Australia on 20 January 1999. His brain had been starved of oxygen during his birth, and he never gained consciousness. His family were heartbroken. His father, Andrew Pearce, was a keen astronomer with an international reputation as a comet observer. He would watch their tails of gas and dust stream through space and marvel as they faded when they moved away from the Sun's warmth, back into a million-year hibernation. But in the hospital, standing next to little Lewis, none of that mattered. His parents held him, spoke to him and kissed him while they awaited the inevitable. But they could not let their son leave the cosmos like this, so one night they wrapped him in a blanket and took him out to show him and talk to him about the stars in the hope that during his brief life he would somehow know something, just something, about the cosmos that he would inhabit so briefly. Like the little girl in Walt Whitman's poem, he might know that there is something.

Asteroid 6916 orbits the Sun somewhere between the orbits of Mars and Jupiter every four years and nine months. It's a large chunk of rock, 12 kilometres wide, and as one of a great many in that region of the solar system it is unremarkable. But as well as a number, it also has a name, Asteroid Lewis Pearce. Long after our Sun has died and spends almost an eternity cooling as a white dwarf, long after the galaxy has merged with others of our local group, this tiny chunk of rock could still be out there. Whatever its fate, and that of the universe, there will always be a portion of space and time that belongs to Lewis Pearce, as it does for you and me. Perhaps that is all that we, and any aliens, can ever have?

I have an inclination and a dream to believe the universe is full of life. I have no proof, but I believe we are not the only creatures, sentient or otherwise, biological or mechanical or both, mortal or eternal, who look at the cosmos in our different ways and differing levels of understanding and wonder what it is we share. As you read this, there are countless alien sunrises taking place across planets in our galaxy and far beyond. The morning skies of alien worlds will come in many forms and may be witnessed by countless strange minds, ceremonies and rituals. Right at this moment, the first and last days of civilisations – billions and billions of them – are taking place with their associated tragedies and triumphs. Somewhere in a distant galaxy a star will go supernova and destroy a billion sentient creatures who have no means of fleeing. Elsewhere, billions of giant worldships silently ply the space between the stars, looking for a new home carrying the libraries of forgotten worlds.

Every year on 1 May, the city of Blaubeuren in Germany organises a walk of about 10 kilometres to the most important caves in the vicinity that have been excavated. At one of them you can see a layer of rocks, dust and ash on which markers have been placed to show the positions of artefacts that have been discovered. One of them is the small ivory plate called the Adorant, with its depiction of the constellation of Orion carved at the start of something that the artist knew nothing about. Perhaps it will be like that if we discover evidence of aliens. Perhaps we humans will always have a middle understanding of the universe and there will be others who are able to see things and grasp concepts we cannot. Or perhaps we and the aliens will need each other as we face the future and whatever might come. There are religions that say God left the universe unfinished for us to complete. Perhaps we will not do that alone?

The composer Gustav Mahler said that we do not compose but that 'we are composed'. We, and all life, are indeed composed by our universe, and I wonder if life will reach its finale someday? Have you listened to the end of Mahler's 9th Symphony? A simple theme meanders away, growing fainter and more distant, but it refuses and refuses to fade, hanging on and on, just another phrase, one more note, and then it diffuses into nothingness. And eternity starts.

Is it that the greatest force in the universe is survival? Returning to Walt Whitman, whose poems are often infused with an almost unbearable sense of loss, he wrote about whispers of heavenly death and the sense that some soul is dying.

But before that happens, imagine that in the faint light of a moon circling a distant world somewhere, something alien is listening to the ocean singing.

INDEX

Q

R

ALSO AVAILABLE

In July 1969, Apollo 11 became the first manned mission to land on the Moon, and Neil Armstrong the first man to step on to its surface. He and his crewmates were the latest men to risk their lives in this extraordinary scientific, engineering and human venture that would come to define the era.

In *Apollo 11: The Inside Story*, David Whitehouse reveals the true drama behind the mission, putting it in the context of the wider space race and telling the story in the words of those who took part – based around exclusive interviews with the key players.

ISBN 978-178578-618-1 (paperback)
ISBN 978-178578-513-9 (ebook)

Nearing half a century since the last Apollo mission, mankind has yet to return to the Moon, but that is about to change. With NASA's Artemis programme scheduled for this decade, astronomer David Whitehouse takes a timely look at what the next 50 years of space exploration have in store.

Based on real-world information, up-to-date scientific findings and a healthy dose of realism, *Space 2069* is a mind-expanding tour of humanity's future in space.

ISBN 978-178578-719-5 (paperback)
ISBN 978-178578-647-1 (ebook)

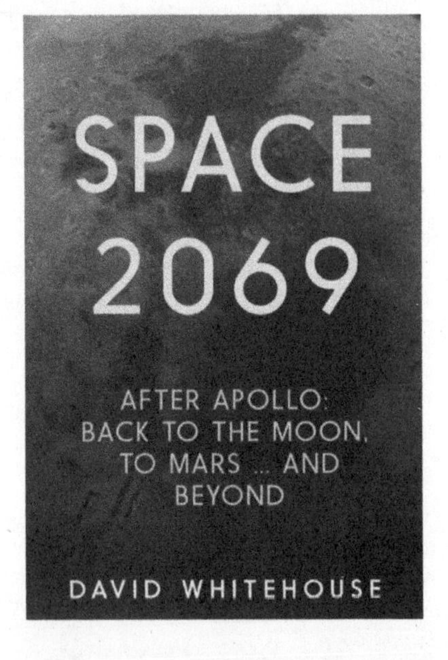